Communications
in Computer and Information Science   686

Commenced Publication in 2007
Founding and Former Series Editors:
Alfredo Cuzzocrea, Dominik Ślęzak, and Xiaokang Yang

More information about this series at http://www.springer.com/series/7899

Jian Cao · Jianxun Liu (Eds.)

Management of Information, Process and Cooperation

Third International Workshop, MiPAC 2016
Hangzhou, China, September 23, 2016
Revised Selected Papers

Editors
Jian Cao
Department of Computer Science
  and Engineering
Shanghai Jiao Tong University
Shanghai
China

Jianxun Liu
School of Computer Science
  and Engineering
Hunan University of Science
  and Technology
Xiangtan
China

ISSN 1865-0929   ISSN 1865-0937   (electronic)
Communications in Computer and Information Science
ISBN 978-981-10-3995-9   ISBN 978-981-10-3996-6   (eBook)
DOI 10.1007/978-981-10-3996-6

Library of Congress Control Number: 2017932643

Printed on acid-free paper

This Springer imprint is published by Springer Nature
The registered company is Springer Nature Singapore Pte Ltd.
The registered company address is: 152 Beach Road, #21-01/04 Gateway East, Singapore 189721, Singapore

Preface

This volume collects the proceedings of the International Workshop on Management of Information, Processes, and Cooperation (MiPAC 2016, formally Process-Aware Systems) held in Hangzhou, China, on September 23, 2016, co-located with the 6th China Conference on Business Process Management (China BPM 2016). Following the success of PAS 2014 and PAS 2015, MiPAC 2016 provided an international forum for exploring challenging research issues in the areas of data/information management, services computing, business processes and workflows, and software engineering, aiming at developing techniques for effective software development to support enterprise applications.

As the third edition in this workshop series, MiPAC 2016 accepted eight qualified papers from 14 submissions. These submissions reported on up-to-date research findings and application case studies.

We would like to thank the Program Committee members for their reviews of the submitted papers. We express our gratitude to the other conference committees, especially to the Program Committee chairs, Jie Wang and Jianxun Liu, and the Steering Committee for their valuable guidance, and to the publicity chair, Cheng Zhang, for his efforts in publishing workshop updates and promoting the workshop in the region. Special thanks to the publication chair, Xiao Liu, for his great efforts, and to the organization chair, Shuiguang Deng, and other staff at Zhejiang University for their attentive preparations for this workshop.

We would also like to take this opportunity to thank the staff at Springer for their efficient work on the publication of the workshop proceedings. Last but not least, we are thankful to the authors of the submissions, the presenters, and all the other workshop participants — the workshop could not be held without their contributions and interest.

February 2017   Jian Cao

Organization

MiPAC 2016 was organized in Hangzhou, China, by Zhejiang University.

Steering Committee

Jianwen Su	UC Santa Barbara, USA
Yun Yang	Swinburne University of Technology, Australia
Jianming Wang	Tsinghua University, China
Liang Zhang	Fudan University, China

General Chair

Jian Cao	Shanghai Jiao Tong University, China

Program Chairs

Jie Wang	Stanford University, USA
Jianxun Liu	Hunan University of Science and Technology, China

Organization Chair

Shuiguang Deng	Zhejiang University, China

Publicity Chair

Cheng Zhang	Anhui University, China

Publication Chairs

Xiao Liu	Deakin University, Australia

Program Committee

Jidong Ge	Nanjing University, China
Jinjun Chen	University of Technology Sydney, Australia
Xiao Liu	Deakin University, Australia
Jie Wang	Stanford University, USA
Cheng Zhang	Anhui University, China
Zhiming Zhao	University of Amsterdam, The Netherlands
Chun Ouyang	Queensland University of Technology, Australia
Jianwei Yin	Zhejiang University, China
Liang Zhang	Fudan University, China

Contents

Process Modeling

Flexible Manufacturing Chain: A SCM for Electronic Commerce Enterprise in Clothing Industry Based on Activiti

Hengheng Wei[1,2], Jidong Ge[1,2(✉)], Chuanyi Li[1,2], Zhongjin Li[1,2], Miaomiao Lei[1,2], and Haiyang Hu[1,2,3]

[1] State Key Laboratory for Novel Software Technology, Nanjing University, Nanjing 210093, China
gjdnju@163.com
[2] Software Institute, Nanjing University, Nanjing 210093, China
[3] School of Computer, Hangzhou Dianzi University, Hangzhou 310018, China

Abstract. With the business development of the process, the manager needs to improve the automation, efficiency and reliability of the business process management, especially in manufacturing area. To solve this problem, we have designed and implemented a supply chain management system, FMC (Flexible Manufacturing Chain), which integrates different aspects of the BPM lifecycle in a unified platform based on Activiti workflow engine. FMC is an effective system to model and execute flexible, data-driven, and user-centric business processes. It enables users to create process models graphically and flexibly, execute process more automatically, monitor the real-time process execution information and improve the process according to the deviations from conformance checking. The FMC system is used as a demo in this article, which is a supply chain system that provides a whole complex product line service including many departments and tasks.

Keywords: Supply chain management · Process modeling · Process execution · Process improvement · Process monitor

1 Introduction

Business Process Management (BPM) is an approach to systematically model, manage and improve business operations inner or among organizations. In BPM, sequences or relationships of business operations are usually described by means of process models that consist of partially ordered control and data flow nodes and can be executed automatically by business process management systems. Process-aware information systems (PAISs) provide support for business processes at the operational level [1, 2] and separate process logic from application code, relying on explicit process models. Participating in the BPM lifecycle requires that each user role is able to create, evolve or execute such process models [3].

FMC (Flexible Manufacturing Chain) is a supply chain management system, which integrates numerous functions and provides an intuitive user interface for the end users

J. Cao and J. Liu (Eds.): MiPAC 2016, CCIS 686, pp. 3–14, 2017.
DOI: 10.1007/978-981-10-3996-6_1

to participate in the entire BPM lifecycle. The system is implemented by the industry standard BPMN [4], the Business Process Model and Notation. Usually, users can execute and control the business process through workflow process engine such as JBPM [5] and Activiti [6], the workflow engine is the core component of FMC system and supply parse, execution and management for BPMN process definitions, FMC chooses Activiti as workflow engine because of its good services, nice integration features with spring framework and its rich log information. However, process execution may deviate from process models due to unexpected happenings or because some employee found a more suitable way achieving the same goal in actual situation. Generally, there are two ways to find deviations between original designed process model and actual running process model, one is process mining and the other is conformance checking.

The basic idea of process mining is to diagnose processes by mining event logs for knowledge [7]. Event logs are the starting point for process mining. The data of the event log can be mined and different aspects about the underlying process can be analyzed. Business process mining, or process mining for short, aims at the automatic construction of models explaining the behavior observed in the event log. Common process mining algorithms discover actual process through the extraction of order relations from event logs, and we can discover deviations from comparison between discovered process and original process.

Conformance checking is applied to measure alignment between designed model and behavior logs, it replays the log event traces on the designed process to evaluate the process in different aspects such as fitness, generalization and precision to find the deviations. FMC workflow system develops and uses process mining and conformance checking plugin on ProM [8] platform, ProM is an open source process mining suit, which can be used for studying process mining cases, and it also integrates a log import tool named XEsame, it can receive different data sources such as database and generate the XLog or XES file which can be used as inputs of conformance checking plugins. We use the BPMN Analysis plugin as conformance checking plugin and develop a χ-algorithm [9] with post-task events plugin on ProM to help improve business process.

With the improve suggestions from process improve module, process designers can modify the process model to a more reliable and efficient version, on the other hand, the accurate and real-time process execution information should also be achieved from process monitor to help users. This system can improve order execution efficiency and reliability as well as save human power.

Section 2 introduces the application scenario used for the demonstration. Section 3 presents the FMC supply chain process management system architecture. Fundamental techniques are described in Sect. 4. Section 5 then describes how the application scenario can be supported with the FMC system. Finally, Sect. 6 concludes the paper.

2   Application Scenario

The case in this article is derived from a China clothes manufacturing company that provides a whole product line service which is very complex and consists of many departments and tasks. Concretely speaking, the services consist of clothes design, material purchase, sample make, volume production, mass customization, storage, delivery etc. Each part of these services includes a lot of task nodes conducted by different departments. FMC utilizes BPMN as process modeling language, we give a description of sample make segment of the whole process model in Fig. 1, we can see there are multiple nodes and different structures in the segment, for example, the verification tasks make concurrent structures respectively, and there is also choice structure, loop structure and sequence structure. These structures compose the whole process and form the process execution logic with the variables in the process model.

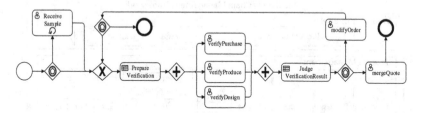

Fig. 1. Sample make BPMN segment.

In the past, the process flow and business logic of the clothes manufacturing process are combined together, designers must spend lots of time and effort to update the process, the process cannot adapt to the change requirement, and sometimes the execution and notification of tasks need to be finished by human power, which is inefficient and unreliable. The clothes manufacturing process execution has a high demand for efficiency, conciseness and reliability.

There should be a graphical interface for designers to create and edit business process easily. The process management and business logic should be separated and the process model should be modified flexibly and better not to restart the process instance. Through structural modeling of business process, the execution of tasks should be judged and executed by process management system automatically to improve process efficiency and reliability, and then the new tasks will be accurately notified to the related users. Multiple process models should be able to be deployed and managed by the system. The users can involve in several workflows at the same time and manage their customized process instances and tasks in a unified manner. Real-time process execution information and statistical information of business processes should be monitored in a respective dashboard, for example, how many process instances are in production status and which is the slowest task in the finished process instances, the execution state of every process instance should be displayed to related users. As well as we know, the deviations will occur compared to the reference model, so the process model should be improved by providing these deviations to model designers and the process instances should be executed more accurately and efficiently.

3   System Architecture

Figure 2 illustrates the major components of the system architecture. In particular, FMC is implemented as a Java EE application utilizing the Activiti BPM engine for process execution, the architecture of FMC consists of three main components including Web User Interface, process management and information management which are respectively discussed in the following sections.

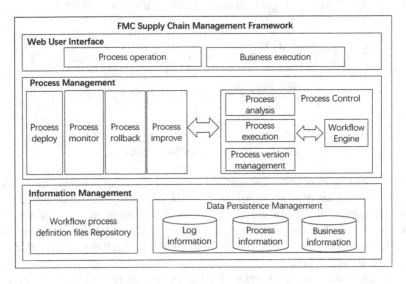

Fig. 2. A SCM system architecture

3.1   Web User Interfaces

Web User Interfaces are based on J2EE web framework. It provides functions for users to execute process operation as well as business operation, receive feedback information graphically. It also enables users to participate in different workflow processes concurrently, so the workflow processes can be shareable between different users after authentication. It interacts with the process management component to deploy, execute, monitor, analyze and improve the workflow processes. We separate the user operations from process operation module and business execution module. The functions of each component are as follows:

- Process operation module: it allows users to create and edit the BPMN 2.0 process model, deploy the process model to the workflow engine in the process management component through web interface, and then the process model can be executed. Users can join the workflow process through authentication, a workflow process can be executed by different users and a user can join different processes. The process operation module also includes process modification to improve the efficiency and simplicity of process model through interaction with the process improve module in the process management component.

- Business execution module: when new tasks are notified to users, users finish the tasks and submit user execution results and process variables to the process control module of the process management component through web interface, and process control module will execute the task and change the process information database.

3.2 Process Management Component

Process Management Component is the core component of the workflow management system architecture, which receives the process inputs from web user interface, and deals with the inputs. Firstly, it can parse process model file definitions and save BPMN model files to database. Secondly, it can receive the execution choices from users, then the workflow engine will analyze the choices with the workflow process information data and update process instances and produce new tasks, then new tasks can be notified to users as well as the monitoring information. Finally, the process can be improved through conformance checking between the logs and the current process models in use, the logs can give expression to the behavior habits of users, then, the improvement suggestions will be provided to model designers to update the process model and make the execution of process more efficient and reliable, the use of BPMN designer makes the modification of process model easily and friendly. The process version management module is used to manage different versions of process definitions with the same process name. The process rollback module is used to roll back the database to previous state when business exceptions or process exceptions are occurred. Through the process management component, the execution and management for business processes can be more automatic, effective, efficient and reliable. This component includes process control module, process monitoring module, process improvement module and process deploy module. The functions of each component are as follows:

- Process control module: it provides the execution, analysis and process version management mechanism to realize the automation and flexibility of business process control. We use Activiti workflow engine to execute the process, and we realize the process analysis sub-module based on different service interfaces such as runtime service interface, history service interface, identify service interface and so on to interact with the core engine to receive user execution results, invoke the core engine to execute and get the information such as new tasks to notify users. Generally, the process management module interacts with the information management component to read and update the process data and is the core module of the workflow management system.
- Process deploy module: it receives the process model definition file from users, then it sends the model files to the workflow engine to check and parse. The parsed process data will be saved to database and the process model definition will be saved in workflow process definition files repository. Then the process can be started to execute.
- Process monitor module: the function of this module is to monitor the process execution in real time. With the execution of the process, users usually want to know statistical information of the process involved in, in the clothes manufacturing

scenario, for example, which orders have finished, aborted, pending or running, and the progress and execution time of related processes. The monitor module needs to interact with the process information in database and return the monitoring information described in graph or table format.

- Process improve module: the module is used to provide improvement suggestions to the model designer. The process models are designed according to the experience of obtained via expert interviews to initially configure a process. During execution however the operational process typically starts deviating from this reference model, for example, due to new regulations that have not been incorporated into the reference model yet, or simply because the reference model is not accurate enough. It implements through process mining plugins or checking conformance between the log recorded in the log information database and the original process model stored in the workflow process definition files repository, and then the deviations are presented to model designer to improve the process model.

3.3   Information Management Component

Information Management Component interacts with process management component, it is designed to store the process-related and business-related data, which includes two main modules, one is workflow process definition file repository which saves the process model definition files deployed to the workflow engine, and another is data persistence management module, this module includes log information sub-module, process information sub-module and business information sub-module. The log information sub-module saves the historic process instances and tasks execution records automatically and these historic process-related tables are applied to find valuable information for process monitoring function. The process information sub-module mainly includes process-related runtime tables such as process instance table, activity instance table, task instance table and variable instance table. The business information sub-module includes business operation data which includes many business entity tables designed by business developers.

4   Fundamental Techniques

4.1   Business Process Modeling Language

The Business Process Model and Notation (BPMN) is an OMG specification that not only defines a standard on how to graphically represent a business process, but now also includes execution semantics for the elements defined, and an XML format on how to store and share process definitions. Process modeling language should contain several key model elements to compose a common process definition. Some key model elements usually used in manufacturing scenario are described as follows:

1. Events: They are used to model the occurrence of a particular event. We use the start event to indicate the start of the process and end events to define the end of the process.

2. Activities: These define the different actions that need to be performed during the execution of the process. Different types of tasks usually are used to describe the model execution such as user tasks, and service tasks. The user tasks are used to represent the business operations, and many related attributes, such as user task id, executor id, variables and judge conditions. The service tasks are used to execute automated operations such as sending emails to customers.
3. Gateways: gateways are used to define multiple paths in the process. Depending on the type of gateway, these might indicate parallel execution, choice. Parallel executions are represented by an inclusive gateway node and converging strategy while choice structures are described by an exclusive gateway node and diverging strategy.
4. Process variables: variable elements define input and output data in the workflow process and they will be used in task nodes and sequence flows which define the different connect relations to judge and confirm execution paths.

We integrate Activiti Modeler to FMC system as open-source web-based editor to create and edit the process models for users. The process models will be saved into workflow process definition file repository so that they can be executed. Figure 1 shows the sample make BPMN segment.

4.2 Process Monitoring Mechanism

When the process is executed, the system is supposed to monitor the status of the process. The system provides monitoring module to gather process status and statistics information according to the customized monitor requirement of different users. For example, the produce managers want to know all running, aborted and finished produce tasks of different process instances in one period time. The structure of process monitor module is described in Fig. 3. Firstly, different users send monitor request through web interface, when servers get requests, they will invoke different service interfaces of process analysis module to compute statistic and real-time process information to users. Finally, the monitoring web page will be updated.

4.3 Process Improvement Mechanism

Process models are used to abstract and structure the business process. With the usage and development of the process model, some deviations will occur compared to the reference model, for example, due to new regulations that have not been incorporated into the reference model yet, or simply because the reference model is not accurate enough [10]. So the users have the demands to improve and update the process models accurately. There are two ways to find deviations between designed process and actual executions. The one way is process mining technique. We implement a process mining algorithm with post-task event trace log as input on ProM, and generate the discovered process model in Petri net formats, then the deviations can be found through the comparison between discovered Petri net and designed model. Another way is conformance checking plugin which takes event logs and designed process model as input and generate the evaluation result in different aspects such as fitness, generalization and

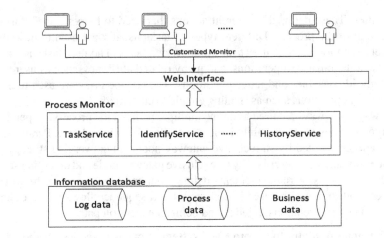

Fig. 3. Process monitor mechanism

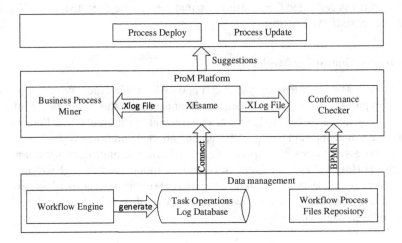

Fig. 4. Process improvement mechanism

precision. With the system running for a period, the task and event log recorded in the database are able to react the actual execution situation, then we use the log format converter to transform the original log files with SQL format to XLog or XES log format that the plugins on ProM [8] can use, then we use the converted log file and process model from repository as inputs to invoke the conformance checker or process mining plugin on ProM. As a result, discovered deviations will be provided to model designers. The structure of the improve module is described in Fig. 4. If model designers modify original process model, the running process instances will be deployed on different versions according to if they can be updated to the new process model.

5   System Demo

5.1   Environment Setting Up

The FMC system is implemented as a J2EE application. The web system integrates spring framework business system, BPMN Designer, MySQL database and Activiti workflow engine. The process management component is developed based on BPMN Designer and Activiti workflow engine. The business execution component is developed as a web system and process and business logs are persisted in MySQL database. And an easy-to-use web interface is designed for process users to interact with server.

5.2   Demonstration Steps

The FMC supply chain management system has been used in a clothes manufacturing company in china. In the manufacturing industry, the technologies and environment are developed and changed very often, to improve efficiency and enhance the market competitiveness. The business processes in use of the manufacturing company often needs re-engineering and improvement. In the past, when the business processes are remodeled, the employees need to understand and learn new rules for a long time. During the time, the running orders have to be executed offline. This problem is solved using the FMC supply chain management system. When the business processes are need to improved, process designers can use process improve module to help provide optimization suggestions and then modify the process models on BPMN designer graphically. With the help of the graphically process designer, process control module and process improvement module, the efficiency and competitiveness of the manufacturing company are improved.

Currently, the system has been applied to manage business processes of clothes manufacturing company. The whole business process functions are shown as follows in Fig. 7. We design the following steps to show the function and the general operating procedures:

1. If a business process will be managed by FMC system, users firstly need to create process models in BPMN2 format with graphical web BPMN process designer. The process will be deployed on workflow engine on server. When the workflow process definition files are received by workflow engine, the engine will parse the files firstly. The nodes with different types will be structured and persisted in the process database.
2. Processes in FMC system are shareable and users can join different processes. When the processes are started, process instances will be created and executed. After users execute business tasks, the execution result will be send to workflow flow engine. The engine analyses and executes related process instances according to the result variables, current process instance state and the relations between task nodes. Then the new tasks will be created through process definition and sent to related users.
3. The process instance has several process states which are Started, Running, Aborted and Finished. Real-time state information of process instances will be collected by process monitor module. this information will be presented to users according to

users' customization with graphs or tables. Figures 5 and 6 show the monitor of running process instances and tasks, which describe the monitoring of process instances and activity instances respectively. Figure 5 shows a list of historic process instances that the user has joined in. The process instances are described from different aspects such as id, time, states so on. Through the process progress button, the Fig. 6 shows the process activities. The yellow boards represent the finished route composed of user tasks and finish time when user tasks are operated. The red boards represent the current tasks for execution. The process monitoring function helps users to get real and important information of executions.

processID	processInstanceID	StartTime	StartActivity	EndTime	EndActivity	State	StartUser	操作
Sample Process	32510	2016-06-26 14:00	StartProcess	2016-06-26 14:07	End	Finished	市场专员 1	过程进度
FMC Process	70001	2016-06-26 20:41	StartProcess			Active	市场专员 2	过程进度
FMC Process	70053	2016-06-26 22:01	StartProcess			Active	市场专员 2	过程进度
FMC Process	70100	2016-06-26 22:25	StartProcess			Active	市场专员 1	过程进度
Sample Process	7501	2016-06-23 19:27	StartProcess			Active	市场主管	过程进度
Sample Process	7544	2016-06-23 19:42	StartProcess			Active	市场主管	过程进度

Fig. 5. Process instance monitor

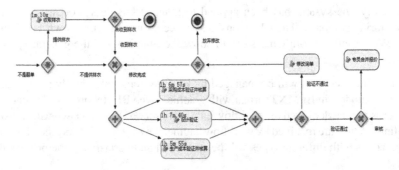

Fig. 6. Activity instance monitor in BPMN graph (Color figure online)

4. When the business process system has running for a period of time, process improvement module allows the business process adaptable to unpredictable changes and can find deviations between logs and models which can provided to process engine as suggestions. We use XEsame on ProM platform to convert process operation records in MySQL database into standard log format which can be used by the conformance checking and process mining plugins on ProM platform. Then the

process engine will invoke the process mining or conformance checking plugins to find deviations and present it to related users. Finally, the process model can be updated to adapt the actual scenario.

5. The system also adopts Spring-Hibernate transaction manager to manage process and business transactions, so the business data and process instance state can roll back to the previous state when the exceptions or the errors happened during the current task execution.

Figure 7 shows the whole supply chain process model. FMC uses BPMN modeling language, the process model contains many complex order relations consist of exclusive gateway, parallel gateway and sequential flows. There are only one start event and many end events in the process. BPMN model language provides a convenient way to create, modify processes for users graphically.

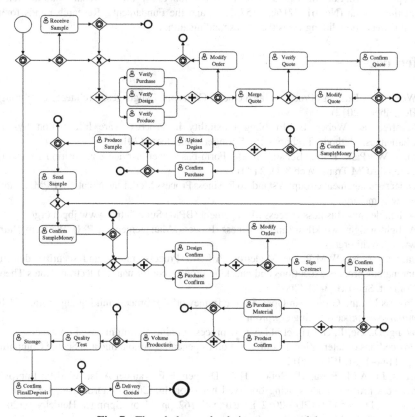

Fig. 7. The whole supply chain process model

6 Conclusion

This paper introduces the FMC supply chain management system from architecture and fundamental techniques aspects, and uses a clothes manufacturing scenario to demonstrate the functions of the system. Through the management of the business process, the users are able to create the structured process models graphically through web interface, to deploy them on server to execute semi-automatically, to monitor the customized real-time process or business execution information, and improve the process according to the deviations from conformance checking and process mining. This system improves the efficiency and reliability of the process execution and management, and allows the process model to adapt to the actual situation accurately and flexibly. A web-based user interface offers an intuitive way to manage process models. The management effect of the clothes manufacturing scenario proves the availability of FMC system.

Acknowledgments. This work was supported by the The National Key Research and Development Program of China (No. 2016YFC0800803), the National Natural Science Foundation, China (No. 61572162, 61572251), and the Fundamental Research Funds for the Central Universities. Jidong Ge is the corresponding author.

References

1. Weske, M.: Business Process Management: Concepts, Languages, Architectures. Springer, Heidelberg (2012)
2. Manfred, R., Weber, B.: Enabling Flexibility in Process-aware Information Systems: Challenges, Methods, Technologies. Springer, Heidelberg (2012)
3. Ingo, W., Paik, H.-Y., Benatallah, B.: Form-based web service composition for domain experts. ACM Trans. Web **8**(1), 2 (2013)
4. Object Management Group: A standard Business Process Model and Notation (BPMN). http://www.bpmn.org/
5. A flexible Java Business Process Management (BPM) Suite. http://www.jbpm.org/
6. A light-weight workflow and Business Process Management (BPM) Platform. http://www.activiti.org/
7. van der Aalst, W.M.P., de Medeiros, A.K.A.: Process mining and security: detecting anomalous process executions and checking process conformance. Electron. Notes Theore. Comput. Sci. **121**, 3–21 (2005)
8. Process Mining Group: An Open Source framework for process mining algorithms (2016). http://www.processmining.org/prom
9. Wang, D., Ge, J., Hu, H., et al.: A new process mining algorithm based on event type. In: Proceedings of International Conference on Dependable, Autonomic and Secure Computing, pp. 1144–1151. IEEE (2011)
10. Buijs, J.C.A.M., Rosa, M., Reijers, H.A., Dongen, B.F., van der Aalst, W.M.P.: Improving business process models using observed behavior. In: Cudre-Mauroux, P., Ceravolo, P., Gašević, D. (eds.) SIMPDA 2012. LNBIP, vol. 162, pp. 44–59. Springer, Heidelberg (2013). doi:10.1007/978-3-642-40919-6_3

Process Enactment

Crowdsourcing Complex Task Automatically by Workflow Technology

Qiang Zheng, Wenyan Wang, Yang Yu[✉], Maolin Pan, and Xiaohui Shi

School of Data and Computer Science,
Sun Yat-Sen University, Guangzhou, China
{zhengq27,wangwy9,shixhui}@mail2.sysu.edu.cn,
{yuy,panml}@mail.sysu.edu.cn

Abstract. Micro-task Crowdsourcing market has become a new trend that convenes a large population of workers to solve the task proposed by a requester online. But these micro-tasks in the market are always simple and independent. To solve complex tasks in real world, some recursive decomposition approaches were proposed and some tools were developed. However, the process of solving complex tasks still involve lots of manual work. How to make this process more automatic? In this paper, we present a new crowdsourcing process model which includes a state machine model of a task and a relation model of tasks. Based on this process model, we design a crowdsourcing platform with the help of state machine workflow technology. With the support of this platform, we can define and execute a crowdsourcing process. In the process of the execution, the platform can manage dependencies between tasks. By means of this platform, one can develop many kinds of crowdsourcing applications with less programming, higher speed and quality. At the end of this paper, a case study is given to demonstrate the practicability of our model and platform.

Keywords: Crowdsourcing · Workflow · State machine · Automation

1 Introduction

Crowdsourcing is a powerful mechanism to complete tasks on the Internet. Recently, crowdsourcing is widely applied in real world. It adopts crowd intelligence and creativity in solving problem. By crowdsourcing, various tasks that are difficult to complete by machine, such as labeling the images, discovering new galaxies (galaxyzoo.org) [1, 2], can be accomplished. These tasks do not acquire any specialized skill. Yet a great deal of tasks demand on specific groups, ranging from crowdsourcing t-shirt designs (Threadless) to software testing (UTest) [3, 4]. Crowdsourcing is also applied in the labor and commercial market to find proper buyers and sellers from different countries (Odesk and Elance) [5].

One of the most popular general-purpose crowdsourcing markets for diverse tasks is Amazon Mechanical Turk (MTurk) and it has been used to research crowdsourcing problems. Tasks on it can vary from images annotation to audio transcription. These tasks, called micro-tasks, are simple, repetitive, independent and cost less time.

© Springer Nature Singapore Pte Ltd. 2017
J. Cao and J. Liu (Eds.): MiPAC 2016, CCIS 686, pp. 17–30, 2017.
DOI: 10.1007/978-981-10-3996-6_2

Compared with typical micro-tasks on the Mechanical Turk, tasks in life are very complex and take a lot of time. They need technicians from different field to work collaboratively. To make this happen in crowdsourcing market, these complex tasks are often decomposed into subtasks. The process of decomposition is very complex. When should it be decomposed and how to break down it are fully decided and executed by crowd workers. This is a completely recursive process with artificial participation.

To automatically solve complex tasks, we propose a crowdsourcing process model on the foundation of state machine workflow technology. In addition to, we illustrate how to design a crowdsourcing platform which supports this model. A variety of crowdsourcing applications can be built quickly by this platform.

2   Related Work

Early, crowdsourcing was used to process large datasets like tagging and classification that were outside the reach of autonomous algorithms. Using crowdsourcing, these tasks can be solved more efficiently, quickly and accurately. But, these tasks don't have any creativity.

Later, more and more problems can be solved via distributed human computation. Soylant provides a word processing interface that uses crowd workers to help with proofreading, document shortening, editing and commenting tasks [6]. VizWiz, an iPhone app which enables blind user to recruit remote sighted workers to help them with visual problems in nearly real-time [7]. Legion is a system that allows end users to easily capture existing GUIs and outsource them for collaborative, real-time control by the crowd [8]. More and more problems of particular type can be solved through crowdsourcing.

Previously, there are three ways to solve complex tasks. TurKit, a toolkit for deploying iterative tasks to MTurk [9]. Its crash-and-rerun programming model that makes TurKit possible, along with a variety of applications for human computation algorithms [10]. The requester must decide how to divide task by hard code before it is posted on MTurk. Crowd-Forge uses the framework of map-reduce to divide complex work into smaller steps [11]. It cannot support iteration or recursion and require the task designer to specify the sequence of activities. The third method is a combination of Price-Divide-Solve algorithm and Turbo-matic tool. The PDS algorithm guides workers through the process of converting large and complex tasks into micro-tasks appropriate for crowd markets [12]. The Turkomatic is a tool for iteration. It can post each step in the best decomposition schemas on Amazon Mechanical Turk [13]. It does not provide crowd workers with the ability to edit workflow. In those toolkits, they rely on MTurk to complete micro-tasks. As a result, they lack the control ability of the whole process. They describe the process of accomplishing complex tasks as workflow, but they do not give business process definition for workflow.

At present, there are two business process modeling standards. One is XPDL defined by WfMC (Workflow Management Coalition). Another is BPMN defined by OMG (Object Management Group). Because the decomposition process of complex task is dynamic, both traditional standard cannot be used for modeling the process of crowdsourcing. Therefore, we use state machine workflow technology to solve this problem.

3 Preliminaries

3.1 State Machine

A finite-state machine (FSM) is a mathematical model of finite computation used to design both computer programs and sequential logic circuits. It is conceived as an abstract machine that has a finite number of states. The machine only stays in one state at a given time. It can transfer from one state to another by a triggering event and condition. This is called a transition. A FSM contains a list of its states, and the triggering condition for each transition. It's a mathematical model can be described as follows. A deterministic finite state machine is a quintuple $(\Sigma, S, s0, \delta, F)$, where Σ is the input alphabet (a finite, non-empty set of symbols); S is a finite, non-empty set of states; $s0$ is an initial state; an element of $S \cdot \delta$ is the state-transition function $\delta: S \times \Sigma \to S \cdot F$ is the set of final states.

Faced with state explosion problems and the lack of ability to describe concurrency, David Harel proposed Statechart diagram [14]. Statechart puts forward concurrent, hierarchy, communication concepts base on FSM and has a better description ability. Later, Object Management Group (OMG) adopted Statechart model to establish UML state machine specification, and World Wide Web Consortium (W3C) published State Chart XML (SCXML) base on Statechart [15, 16]. They have some common concepts, for example State, Composite State, Event, Condition, Action and Transition. The state refers to a situation of object in its life cycle. That an object stay in some particular state is bound to meet certain conditions and wait for some events. Transition is a kind of relationship between two states, when a certain event occurs and satisfies specific guard condition, source state will transfer to the target state with a certain action. A transition in diagram often expressed like this: Event [Condition]/Action.

3.2 State Machine and Workflow

The state machine is widely used in many fields. In software design field, UML state machine diagram are used to model lifecycle and dynamic behaviors of an object or system. In the course of business process, state diagram as a modeling approach are used to build workflow model. There are many scholars introduce how to use Statechart modeling workflow. Yang Dong couples state with activity and expand the types of events in Statechart [17]. Wai Yin Mok make an analogy between four common workflow patterns and some concepts in Statechart and point out three properties model should satisfy: Termination, Confluence, Observable Determinism [18]. Kushnareva use Statechart to model Crisis Management and solve the problem of the process of dynamic change [19]. Guy Redding adds region and gateway to expand the state machine to support flexible process model [20]. State machine diagram is also used to model the lifecycle of business artifact in Artifact-Centric process model [21]. More than others, wendy dwell on the possibility of Statechart as workflow Specification [22]. Again, in a engineering application, windows workflow foundation framework and OSWorkflow engine support using state machine to establish workflow model [23, 24]. Here, using state machines diagrams modeling workflow model is called state machine workflow technology.

4   Modeling

In order to solve complex task automatically on single platform without the support of other third-party tools, the following modelling method which combines the PDS algorithm and state machine workflow technology is proposed.

4.1   Conceptual Meta-model of Crowdsourcing

To define and introduce the model, some basic concepts are defined as following:

Task: task is the question posted by requester. It can be classified into simple task and complex task.

Simple task (STask): These tasks can be solved by worker directly and don't need to be divided. It is also equivalent to the micro-task.

Complex task (CTask): Tasks should be decomposed into several parallel subtasks. These subtasks are CTask or STask.

Figure 1 is conceptual meta-model of crowdsourcing which shows the relationship between CTask and STask. Round box represents CTask and ellipse represents STask. A directed edge $e_{a,b}$ implies that task B is one of subtasks of task A. The solution of task A is composed of B and C's solution.

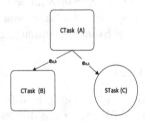

Fig. 1. Conceptual meta-model

4.2   Modeling Life Cycle of Crowdsourcing Task Object

According to object-oriented thinking, a task can be regard as an object. The task will experience a series of steps until it gets solution. Each step can be considered as what needs to be done in a particular state. These states can be expressed by object lifecycle. The life cycle of task object can be varied. Figure 2 presents a kind of lifecycle of task by State Machine Diagram. It contains the following basic state S = {Initial, Judging, Decomposing, DecomposeVoting, Waiting, Solving, SolveVoting, Merging, Final}. The Initial and Final state are pseudo-states. Waiting state is a flag state. Each of the remaining state corresponds to a certain number of work items, which are performed by workers or service. Relevant states in our model are explained as following.

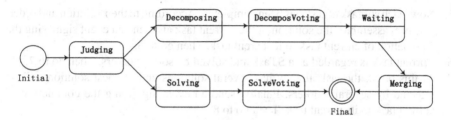

Fig. 2. The lifecycle of crowdsourcing task

Judging state: The current task is undergoing artificial judge phase. It requires a certain number of people to judge whether current task is complex. The definition of complex task should be determined by crowd experts.

Decomposing state: The current task is complex and it requires some workers to decompose it into smaller tasks.

DecomposeVoting state: The current task already has several decomposition schemas. It needs some workers to vote on which is the most reasonable and reliable one.

Solving state: The current task is simple and it requires a certain number of workers to solve it.

SolveVoting state: The current task has been solved by several workers. It is time to vote for the best solution.

Waiting state: It is a flag state and the only function is to wait on those events that are delivered when subtasks get their solution.

Merging state: The current task is complex and its all subtasks have gotten their solution. All the subtasks' solutions are assembled into solution of the current task by crowd workers or service task.

4.3 Crowdsourcing Process Modelling

During task execution, crowdsourcing process model is generated automatically according to meta-model and task's lifecycle model. In state machine workflow, state machine diagram is process definition of workflow instance. If a requester issues a task A, a new workflow instance corresponding to task A will be started. This instance will go through such a process.

1. Determine whether or not task associated with this instance is complex in Judging state, if the task is a CTask go to 2, otherwise go to 6.
2. Each of workers in this step gives a decomposition strategy of task associated with this instance. The decomposition strategy is made up of several small steps. Then go to 3.
3. Several decomposition solutions were produced. They will be voted by several workers to obtain the best decomposition solution. Each step in the best decomposition scheme will be released to the platform as a subtask of current task. Each subtask has its own workflow instance. For each subtask, go to 1. For current task, go to 4.
4. Wait for each subtask generated in 3 to send an event. The event means current subtask has been done. Thereafter, go to 5.

5. Now, each subtask of task has been completed. According to their solution and order, platform assembles the solution of the current task and send an event signifying the completion of current task to its parent task. Then go to 8.
6. Current task is regarded as a STask and solved by some workers. Then go to 7.
7. At this point, the task already has several solutions and the best solution will be selected by several workers. Finally, send an event signifying the completion of current task to its parent task. Then go to 8.
8. Now, current task is accomplished.

Among these steps, the tasks which experience a step sequence including 1, 2, 3, 4, 5, 8 have a process of decomposing and merging. The Relationships between tasks and subtasks are recorded during decomposition. Finally, it is used to complete merging process.

Figure 3 shows the lifecycle of task A. The dotted line connects the task and its lifecycle. The red state indicates workflow instance have gone through, while green state means workflow instance stays in it. Task A has gone through Judging, Decomposing, DecomposeVoting state. The best decomposition strategy has been found in DecomoseVoting state.

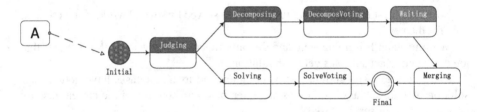

Fig. 3. Lifecycle of task A (Color figure online)

When workflow instance exits DecomposeVoting state, the exit action regards each step of the decomposition strategy as a subtask and launch a corresponding workflow instance. In this case, if the best decomposition policy of task A are B, C and D, then task A enters Waiting state to wait task B, C, D to be solved. Figure 4 express the status after exiting the DecomposeVoting.

Once subtasks B, C, D begin their life cycles, they also need to be judged whether or not it is a complex task. When C is judged to be a complex task and B, D are judged to be simple tasks. B and D are solved by workers directly. Then C experiences the same life cycle as A and generates two subtasks E, F. According to the concept of meta-model and decomposition process, it gradually forms a tree structure. The left side in Fig. 5 is a tree generated in decomposition process.

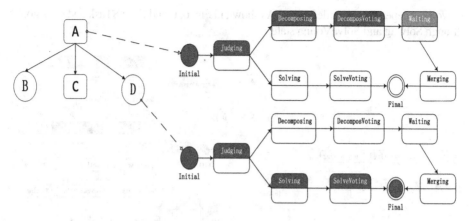

Fig. 4. A recursive process of task A

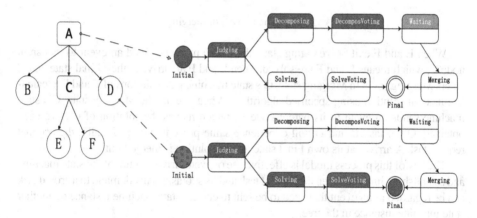

Fig. 5. A tree generated in decomposition process

As can be seen from the decomposition process, the tree generated by it has the following characteristics.

1. Leaf nodes correspond to simple tasks and all non-leaf nodes are complex tasks.
2. Node with one degree doesn't exist.
3. The height and degree of the tree are completely determined by the decomposition process with artificial participation.

Each state machine instance in tree must hold same reference to this tree. When current task is divided into a series of subtasks, the state machine instance of each subtask is inserted into the tree as a child of current node. Each subtask judged to be Complex task repeats the above process, then decomposition tree is built. This tree records the decomposition relationship between tasks. With the tree, the merging process begin to perform.

Once the task is broken down, it will wait for its subtasks to be solved and send event to it. Finally, STask is solved by workers directly and CTask is solved by merging algorithm or workers depending your definition. As shown in Fig. 6, E and F are STask and have gone through Solving and SolveVoting state.

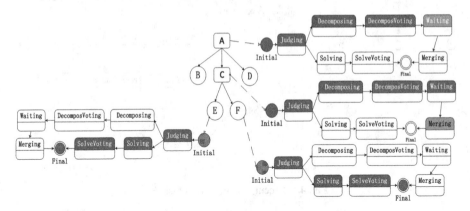

Fig. 6. The process of merging

When E and F exit SolveVoting state, their state machines send an event to its parent node C, which means E and F are solved. Then E and F all arrive at their Final state and C enters Merging state. In Merging state, C's state machine gets solutions of E and F to calculate its own solution using specified algorithm. When C exits the Merging state, C's state machine sends an event to its parent node A, which means the solution of C have been obtained. Other task B and D will experience same procedure of E. Finally, the original request task A arrives at its own Final state and its solution is merged out.

The key of this process model is effective use of event-driven nature of the state machine and establishing links between the current task and its subtasks. Any demand to a crowd task can be regarded as an event that need to be sent to current state machine instance or another state machine instance in the tree.

5   Model Implementation

This section focuses on how to make crowdsourcing process Model into reality. As mentioned in the previous section, state change process of object can be described by a state machine, so a state machine standard and its implementation are needed. SCXML is a W3C recommendation standard of state machine. Its transition is certain. Apache Commons SCXML is an implementation of SCXML standard by java language. We develop a tool called BOWorkflow based on Commons SCXML [25, 26]. It supports crowdsourcing process model mentioned above.

5.1   Commons SCXML Introduction

Commons SCXML is an implementation aimed at creating and maintaining a Java SCXML engine capable of executing a state machine defined using a SCXML document, while abstracting out the environment interfaces. As shown in Fig. 7, Commons SCXML provides abilities of the following.

1. SCXML Parser: parse scxml document.
2. SCXML Data Mode: support some script language.
3. Content and Evaluators: provides expression evaluation and application context environment.
4. SCXML Executor: The switch of state machine.
5. Triggering Event: An abstract about events.
6. Custom Actions: support custom action in program.
7. Custom Semantics: support custom execution semantics of state machine.

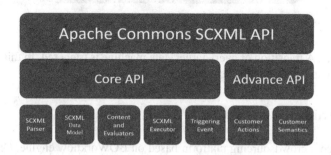

Fig. 7. The architecture of Commons SCXML

Obviously, the Commons SCXML only provides the runtime environment and executive capability of single state machine. It cannot effectively handle communications between multiple state machines. To support crowdsourcing process model, Commons SCXML need to be extended.

5.2 Total Design of BOWorkflow

Here introduce how to integrate these features mentioned above with Commons SCXML. The Runtime Environment of BOWorkflow as shown in Fig. 8.

Entire runtime environment is composed of workflow engine and all state machine instances. The workflow engine is made up of SCXML parser, task dispatcher, instance manager, multi state machine event dispatcher and custom semantic components. SCXML parser is in charge of parsing state machine definition. Task dispatcher is responsible for assigning work items which is defined in a state. Instance manager provides the runtime information of all workflow instances. Multi State Machines Event Dispatcher delivers events from inside or outside to other state machine instances. A state machine instance can directly respond to external events through SCXML Executor. Sending process for internal events is that current state machine instance call event dispatcher to deliver event to target. The dispatcher gets the identifier of target state machines by parsing the tree of source state machine instance, then get the reference of target state machine instances by Instance Manager. At last, dispatcher forward this

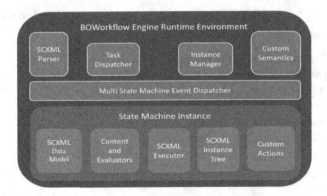

Fig. 8. The runtime of BOWorkflow

event to target. A transition of every target state machine is triggered. In this transition, some action can be executed. These actions can be varied.

6   Case Study

We developed a crowdsourcing platform based on BOWorkflow engine. On this platform, the crowdsourcing developer only need to focus on developing human-machine interface, designing good state machine diagram and writing proper merging algorithm. Some basic state machine diagrams are built-in. Finally, we made a case study based on this Platform to demonstrate the feasibility of the model. The case study is to write an article about crowdsourcing.

Firstly, crowdsourcing platform defines a state machine diagram as shown in Fig. 9. It represents crowdsourcing task object's lifecycle. The action defined on entry will be executed when instance entered a state. The "do" is inner transition. A transition in diagram often expressed as: Event [Condition]/Action. In some states, engine needs to execute UserTask action to assign a work item to a group of workers. For simplicity, some strategies and explanations are specified as follows.

1. In Decomposing state, engine will insert a work item into two workers' work list for decomposing current task.
2. In DecomposeVoting and SolveVoting state, engine will insert a work item task into three workers' work list for voting current task.
3. UserTask: An action to assign work item to workers or an individual one.
4. Count: it is an action to count the number of workers who deem current task as a simple task.
5. GetBestDec: An action to get best decomposition schema based on the vote.
6. GetBestSol: An action to get best solution based on the vote.
7. Merge: A merging algorithm which is used to merge into the solution of the current task on the basis of the solutions of subtasks. The algorithms for merging different types of tasks are different. For example, to write an article, the merging algorithm.

8. NewStateMachine: An action to create a sub-state-machine which corresponding a step in best decomposition schema and insert its identifier into the decomposition tree as child node of current state machine.
9. SendToSelf: Send an event to the internal event queue of current state machine.
10. SendToParent: Send an event to the external event queue of parent state machine.

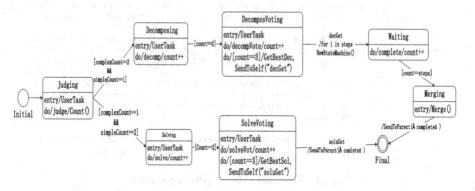

Fig. 9.　The lifecycle of crowdsourcing task

In the end, this task was decomposed into a set of subtasks as shown in Fig. 10. All solutions of simple tasks are merged into the solution of original request task as shown in Fig. 11. In this case, the merging algorithm use a '</br>' which is a line separator in webpage to splice subtasks' solutions.

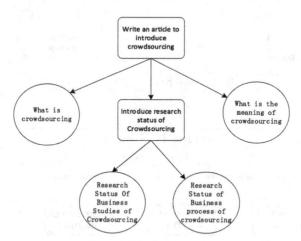

Fig. 10.　Decomposition result of this case study

As shown in Fig. 12. Requester posts a task, then engine create a new crowdsourcing task workflow instance. After initialization, this instance enters a state. In this state, engine dispatches a work item to worker. When worker complete it, worker send an event to tell engine work item has been finished. After received this event, engine

Solution	
Task Name	Write an article to introduce crowdsourcing
Task Description	it is necessary to introduce the origin of crowdsourcing, research status and the significance of crowdsourcing
Task Solution	Crowdsourcing, a modern business term coined in 2006, is defined by Merriam-Webster as the process of obtaining needed services, ideas, or content by soliciting contributions from a large group of people, especially an online community, rather than from employees or suppliers
	Crowdsourcing business model including online labor service, crowdsourcing translation, crowdsourcing writing, etc. There are more famous company, such as Amazon, CrowdFlower, TopCoder and so on.
	The crowdsourcing research on academic mainly includes workflow, task assignment, hierarchy, real-time response, synchronous collaboration, quality control, reputation, and motivation. Most of the research is still in its infancy.
	There is great significance. For people, it liberates people's work time. For enterprise, it equivalent to hired the most professional workers

Fig. 11. The result of the original request task

forward it to the corresponding workflow instance. Then workflow instance executes a transition as response. In this transition, some predefined actions are executed, such as SendToParent, NewStateMachine and so on. It is with these actions; the task can communicate with other tasks in its process of completing. So workers can work together to complete a certain goal.

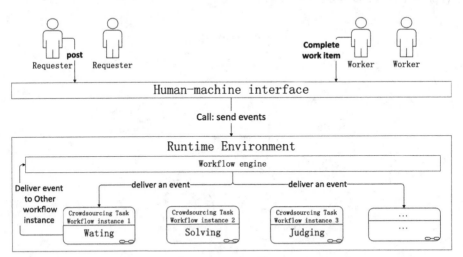

Fig. 12. Interaction diagram of user and the platform

7   Conclusion and Future

This article discusses how to build a general-purpose crowdsourcing platform which support complex tasks based on state machine workflow technology. This platform makes the process of decomposition, resolving, and merging of complex task more automated. It mainly links task object and state machine together. At the same time, give more responsibility to Action, enable it not only record the relationships between tasks and subtasks in the process of decomposition, but also send events to any state machine based on previous records. Based on crowdsourcing platform that implements this model, crowdsourcing application can be built quickly. Crowdsourcing application developers only need to focus on developing human-machine interface, designing good state machine diagram and writing proper merging algorithm. In future work, we plan to concentrate on improving workflow engine and developing visual modeling tools which support this model, and integrate scientific workflow in the process of merging.

Acknowledgements.   This work is Supported by the National Natural Science Foundation of China under Grant No. 61572539; the Research Foundation of Science and Technology Major Project in Guangdong Province under Grant Nos. 2015B010106007, 2016B010110003; the Cooperation Project in Industry, Education and Research of Guangdong Province & Ministry of Education of China under Grant No. 2013B090500103; the Research Foundation of Science and Technology Plan Project in Guangdong Province under Grant No. 2016B050502006; the Research Foundation of Science and Technology Plan Project in Guangzhou City under Grant No. 2016201604030001.

References

1. Yuen, M.C., King, I., Leung, K.S.: A survey of crowdsourcing systems. In: Privacy, Security, Risk and Trust (PASSAT) and 3rd IEEE International Conference on Social Computing (SocialCom), pp. 766–773. IEEE Press, Boston (2011). doi:10.1109/PASSAT/SocialCom.2011.203
2. Kittur, A.: Crowdsourcing, collaboration and creativity. J. ACM Crossroads **17**, 22–26 (2010). doi:10.1145/1869086.1869096
3. Brabham, D.C.: Moving the crowd at threadless: motivations for participation in a crowdsourcing application. J. Inform. Commun. Soc. **13**, 1122–1145 (2010). doi:10.1080/13691181003624090
4. Vukovic, M.: Crowdsourcing for enterprises. In: Congress on Services-I, Los Angeles, pp. 686–692 (2009). doi:10.1109/SERVICES-I.2009.56
5. Doan, A., Ramakrishnan, R., Halevy, A.Y.: Crowdsourcing systems on the world-wide web. J. Commun. ACM **54**, 86–96 (2011). doi:10.1145/1924421.1924442
6. Bernstein, M.S., Little, G., Miller, R.C., et al.: Soylent: a word processor with a crowd inside. J. Commun. ACM **58**, 85–94 (2015). doi:10.1145/2791285
7. Bigham, J.P., Jayant, C., Ji, H., et al.: VizWiz: nearly real-time answers to visual questions. In: Proceeding of the 23rd Annual ACM Symposium on User Interface Software and Technology, pp. 333–342. ACM Press, New York (2010). doi:10.1145/1866029.1866080
8. Lasecki, W.S., Murray, K.I., White, S., et al.: Real-time crowd control of existing interfaces. In: Proceedings of the 24th Annual ACM Symposium on User Interface Software and Technology, pp. 23–32. ACM Press, Santa Barbara (2011). doi:10.1145/2047196.2047200

9. Little, G., Chilton, L.B., Goldman, M., et al.: Turkit: tools for iterative tasks on mechanical turk. In: Proceedings of the ACM SIGKDD Workshop on Human Computation, pp. 29–30. ACM Press, Washington (2010). doi:10.1145/1600150.1600159

10. Little, G., Chilton, L.B., Goldman, M., et al.: Turkit: human computation algorithms on mechanical turk. In: Proceeding of the 23rd Annual ACM Symposium on User Interface Software and Technology, pp. 57–66. ACM Press, New York (2010). doi:10.1145/1866029.1866040

11. Kittur, A., Smus, B., Khamkar, S., et al.: Crowdforge: crowdsourcing complex work. In: Proceedings of the 24th Annual ACM Symposium on User Interface Software and Technology, pp. 43–52. ACM Press, Santa Barbara (2011). doi:10.1145/2047196.2047202

12. Kulkarni, A., Can, M., Hartmann, B.: Collaboratively crowdsourcing workflows with turkomatic. In: Proceedings of the ACM 2012 Conference on Computer Supported Cooperative Work, pp. 1003–1012. ACM Press, Seattle (2012). doi:10.1145/2145204.2145354

13. Kulkarni, A.P., Can, M., Hartmann, B.: Turkomatic: automatic recursive task and workflow design for mechanical turk. In: CHI 2011 Extended Abstracts on Human Factors in Computing Systems, pp. 2053–2058. ACM Press, Vancouver (2011). doi:10.1145/1979742.1979865

14. Harel, D.: Statecharts: a visual formalism for complex systems. J. Sci. Comput. Program. **8**, 231–274 (1987). doi:10.1016/0167-6423(87)90035-9

15. UML State Machine. http://www.omg.org/spec/UML/2.5/

16. SCXML. https://www.w3.org/TR/scxml/

17. Dong, Y., Shensheng, Z.: Modeling workflow process models with statechart. In: Proceedings of 10th IEEE International Conference and Workshop on the Engineering of Computer-Based Systems, pp. 55–61. IEEE Press, Huntsville (2003). doi:10.1109/ECBS.2003.1194783

18. Mok, W.Y.: Revisiting Workflow modeling with statecharts. J. Adv. Top. Database Res. **3**, 237–256 (2003). doi:10.4018/978-1-59140-255-8.ch012

19. Kushnareva, E., Rychkova, I., Grand, B.: Modeling and animation of crisis management process with statecharts. In: Matulevičius, R., Dumas, M. (eds.) BIR 2015. LNBIP, vol. 229, pp. 145–160. Springer, Heidelberg (2015). doi:10.1007/978-3-319-21915-8_10

20. Redding, G., Dumas, M., Hofstede, A.H.M., et al.: Generating business process models from object behavior models. J. Inform. Syst. Manage. **25**, 319–331 (2008). doi: 10.1080/10580530802384324

21. Gerede, C.E., Bhattacharya, K., Su, J.: Static analysis of business artifact-centric operational models. In: IEEE International Conference on Service-Oriented Computing and Applications (SOCA 2007), pp. 133–140. IEEE Press. Newport Beach (2007). doi:10.1109/SOCA.2007.42

22. Zhang, W.W., Beaubouef, T., Ye, H.: Statechart: a visual language for workflow specification. Int. J. Comput. Theor. Eng. **4**, 921 (2012). doi:10.7763/IJCTE.2012.V4.607

23. State Machine Workflow in Windows Workflow Foundation. https://msdn.microsoft.com/enus/library/ee264171(v=vs.110).aspx

24. OSworkflow. https://java.net/projects/osworkflow

25. BOWorkflow. https://github.com/ThinerZQ/BOWorkflow

26. Apache Commons SCXML. http://commons.apache.org/proper/commons-scxml/

An Adaptive Scheduling Mechanism for Analytical Workflow Model

Yan Yao and Jian Cao[✉]

Shanghai Jiao Tong University, Shanghai, China
{yaoyan, cao-jian}@sjtu.edu.cn

Abstract. With the recent development of big data and cloud computing, more and more applications of data analytics emerged. Cloud workflow is a good tool to orchestrate analytical tasks, which is called analytical workflow. In this paper, we focus on the resource scheduling for analytical workflow. As there exist multiple instance for each task when executing, the execution time of workflow is dynamical change with the resource. First of all, we model the performance of analytical workflows executed in cloud and formulate the scheduling problem that minimizing the execution time with budget constraint. Then, we propose an adaptive scheduling algorithm and take machine learning algorithm as case study to illustrate the performance of our algorithm.

Keywords: Big data analytic · Analytical workflow · Adaptive scheduling · Cloud computing

1 Introduction

Today, more and more organizations are collecting, storing, and analyzing massive amounts of data, which is generated from many sources (web sites, social media, sensors, etc.). This data is commonly referred to as "big data" because of its '4 V' feature, which is volume, velocity, variety and value [1]. People are realizing the potential value of this data and using it to identify new opportunities. A key to deriving value from big data is the use of analytics – big data analytic, which will convert data into information for decision-making by users.

Big data analytic allows users to make better and faster decisions. Typical process of data analysis can be divided into several phases. Data are assessed and selected, cleaned and filtered, visualized and analyzed, and the analysis results are finally interpreted and evaluated [2]. A way to realize the big data analysis is constructing and orchestrating the analytical tasks using workflow, which can make data analysis agile. We call the workflow for big data analysis as analytical workflow. Analytical workflows composed of atomic analytic components for data selection, feature extraction, modeling, and scoring.

The cloud is now in the mainstream of computing. With the cloud computing, resources are virtualized and offered as a service over the Internet. The potential benefits of the cloud include access to specialized resources, quick deployment, easily expanded capacity, the ability to discontinue a cloud service when it is no longer needed and cost savings. These same benefits make the cloud attractive for analytical workflows.

© Springer Nature Singapore Pte Ltd. 2017
J. Cao and J. Liu (Eds.): MiPAC 2016, CCIS 686, pp. 31–45, 2017.
DOI: 10.1007/978-981-10-3996-6_3

The current cloud services are available as Infrastructure as a Service (IaaS), Platform as a Service (PaaS), and Software as a Service (SaaS) [3]. IaaS clouds provide virtualized hardware and storage for users to deploy their own applications, and therefore are most suitable for executing analytical workflows.

Real-world IaaS cloud services such as Amazon EC2 [4], provide resources in the form of virtual machine (VM for short) instances to meet different demands of various applications. Virtual machines are usually charged by the provisioned time units, such as minutes or hours. Within the same cloud, VMs work in a structure as a virtual cluster and data transfers are typically performed through a shared storage system without financial charge; while across different clouds, users generally need to pay for inter-cloud data transfers.

Due to the nature of cloud computing that makes computing a utility, one major objective of resource provisioning in clouds is to allocate thus pay for only those cloud resources that are truly needed. Because that individual components may have multiple instances when processed by workflow engine, which will result in the dynamical execution time. Because that the execution time of the components will depended on the number of virtual machines. That is, the execution time of each component is uncertain before executed and thus the resource scheduling for analytical workflows will be much more complex. In this paper, we construct mathematical models to quantify the performance of analytical workflows in IaaS cloud, and formulate a task scheduling problem to minimize the workflow execution time under budget cost constraint. We then design an adaptive scheduling mechanism to this problem. And we take *KNN* (K Nearest Neighborhood algorithm) as a case study to illustrate the performance of our algorithm.

The rest of the paper is organized as follows. Section 2 conducts a survey o related workflow on workflow optimization especially in cloud environments. Section 3 models the performance of analytical workflows executed in cloud and formulated the scheduling problem that minimizing the execution time with budget constraint. An adaptive scheduling algorithm is given in Sect. 4. Section 5 presents simulation result. Finally, the conclusion is given in Sect. 6.

2 Related Work

We conduct a simple survey of related work on workflow scheduling in cloud environments [5–7]. To schedule workflow in cloud computing environment, application schedulers may have different objectives, including minimizing total execution time, minimizing total monetary cost, balancing the load among resources, and achieving stable performance, etc. We classify the workflow scheduling in cloud into two categories: The best-effort scheduling and QoS-constrained workflow scheduling.

The best-effort scheduling attempts to optimize one objective while ignoring other factors such as various QoS requirements (e.g., time, cost, and reliability) [7, 10, 11, 14]. In contrast to best-effort scheduling, QoS-constrained workflow scheduling is more close to real-world applications. A QoS-constrained schedule tries to optimize some objective with constraints on other objectives. Among all the QoS-constrained workflow scheduling problems for cloud system, deadline-constrained and budget-constrained workflow scheduling are two primary categories that are widely studied in the literature.

In this paper, we focus on the budget constrained scheduling for analytical workflow. The intuition of this problem is to finish a workflow as fast as possible at given budget. Existing algorithms on budget-constrained workflow scheduling have two categories: One-time heuristic algorithm [8–10] and back-tracking-based heuristic algorithm [12–14]. In general, most of one-time heuristic algorithm are extended from HEFT [11], which is a typical scheduling algorithm in grid computing. Thus, they have high time complexity. The basic idea of back-tracking heuristic is to start from an assignment which has good performance under one of the two optimization criteria considered (that is, makespan and budget) and swap tasks between resources trying to optimize as much as possible for the other criterion.

The most distinct feature that makes the scheduling for analytical workflow different from other general workflows in cloud is multiple-instances. For analytical workflows, some components may generate multiple instances (can be view bag of tasks) when interpreted by workflow engine. This will result in dynamical execution time depending on the resources. Hence, new workflow scheduling methods should be developed for analytical workflow.

3   Mathematical Models and Problem Formulation

As illustrated in Fig. 1, there are three layers in the analytical workflow scheduling problem in cloud environments: the workflow graph layer comprises of interdependent workflow tasks, the virtual machine layer representing a network of virtual machines, and IaaS cloud layer which hosting virtual machines. We consider a one-to-one mapping scheme such that each task in the workflow is assigned to a different virtual machine for execution. However, in practice, once the scheduling is obtained, we can always reuse the virtual machines to reduce the actual number of virtual machines being created and used.

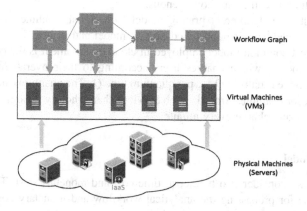

Fig. 1. Workflow execution in IaaS cloud

3.1   Analytical Workflow Model

Generally, a analytical workflow is modeled as a weighted directed acyclic graph (DAG) $A_c(V_c, E_c)$, V_c means components and edge e_{ij} representing the data dependency between component c_i and c_j. The weight on node t_i is the execution time of the component. More formally, we denote its earliest start time and earliest finish time as $EST(c_i)$ and $EFT(c_i)$ respectively, and denote its latest start time and latest finish time as $LST(c_i)$ and $LFT(c_i)$, respectively, $c_i \in V_c$. The buffer time of component c_i defined as $LST(c_i) - EST(c_i)$ or $LFT(c_i) - EFT(c_i)$, is the amount of processing time that component c_i can be delayed without affecting the overall execution time of the entire workflow [12].

$$EST(c_i) = \begin{cases} 0 & \text{if } c_i \text{ is entry node} \\ max_{c_p \in P_i}\{EST(c_p) + t_i\} & \text{otherwise} \end{cases} \tag{1}$$

$$EFT(c_i) = EST(c_i) + t_i \tag{2}$$

$$LST(c_i) = LFT(c_i) - t_i \tag{3}$$

$$LFT(c_i) = \begin{cases} EFT(c_i) & \text{if } c_i \text{ is exit node} \\ min_{c_p \in P_i}\{LST(c_p)\} & \text{otherwise} \end{cases} \tag{4}$$

For some components of an analytical workflow, multiple instances (tasks) may be generated when the workflow model is executed by workflow engines. We assume the instance execution time of the components are either known or can be obtained using profiling and performance estimation techniques.

3.2   Cloud Resource Model

Resources are offered in the form of virtual machines by cloud providers, such as Amazon EC2 [15], Microsoft Azure [16] and Google Compute Engine [17]. In practice, a cloud platform may provide various type virtual machines, each with different capacity and price. For simplicity, we consider only one type of virtual machine, which means the virtual machines are homogenous.

There exist several different pricing models for virtual machines, consumption-based pricing model, subscription-based pricing model and market-dependent pricing model. The most common model employed in cloud environment is the consumption-based pricing model, which charges users according to their overall resource consumption. The price usually is Q_{price} per quantum time Q_t. The quantum can be one hour or one minute, for example, Amazon Web Services charge hourly, Google App Engine and Windows Azure charge every minute.

3.3   Cost Model

In this paper, we consider two type cost: time cost and monetary cost. The time cost means the time for processing the analytical workflow and monetary cost means the cost for renting virtual machines in cloud.

We consider a set of m components to be executed and each component with k_i instance $e_{i,j}$, $j \in [1, k_i]$. For a component c_i, the execution time of all its instances are same, denoted as t_i. The execution time T_i of component e_i on a virtual machine is calculated as:

$$T_i = T(I) + T(E_i) + T(R_i) \tag{5}$$

where $T(I)$ denotes the star up time of a virtual machine, $T(E_i) = ceil(k_i/n_i) \cdot t_i$ denotes the time for running component c_i (n_i is the number of allocated virtual machines, ceil() is a rounded up function), and $T(R_i)$ denotes the time of downloading and uploading data from cloud storage system to the virtual machine. The upper bound of $T(E_i)$ is $k_i * t_i$ and lower bound is t_i. The time duration T_i spans from the star up of the virtual machine to the end of output data transfer from c_i.

Similarly, the monetary cost C_i of executing a specific component a_i, $i \in \{1, 2, \ldots, m\}$ for a duration of T_i is the sum of three cost parts:

$$C_i = C(compute) + C(transfer) + C(storage) \tag{6}$$

where $C(compute)$ denotes the cost of renting a virtual machine, $C(transfer)$ denotes the cost of transferring the data required and produced by the component, and $C(storage)$ denotes the data storage cost of a_i. In most of cloud providers, e.g. Amazon S3, Google and Azure, the data transfer in the same cloud are free, and storage are monthly charged. Therefore, we only consider compute cost $C(compute)$.

$$C_i = \begin{cases} n_i \cdot T(E_i)' \cdot Q_{price} & \text{if } \varphi_i = 0 \\ (n_i \cdot T(E_i)' + (n_i - \varphi_i) \cdot t_i') \cdot Q_{price} & \text{if } \varphi_i \neq 0 \end{cases} \tag{7}$$

where $T(E_i)' = ceil(T(E_i))$ is the time rounded toward positive infinity of $T(E_i)$, $t_i' = round(t)$ is the time rounded toward nearest integer, and $\varphi_i = mod(k_i/n_i)$ is the remainder after division.

3.4   Problem Formulation

Based on the mathematical models constructed above, we formally formulate the analytical workflow scheduling problem for minimum execution time under a used-specified cost constraint in cloud environments.

Given an analytical workflow graph $G_c(V_c, E_c)$, we first provide the definition of *critical path* (CP) of a workflow graph.

Definition. *Critical Path (CP)*: the longest path in the workflow graph weighted with time cost, which consists of all the components with zero buffer time.

In addition, we consider several constraints on the mapping between workflow and cloud resource as follows.

(1) Each virtual machine can execute only one component at a time.
(2) An analytical component cannot start execution until all its required data arrived.
(3) A dependency edge cannot start data transfer until its proceedings finished.

Given a fixed financial budget B, we wish to find how many that the virtual machines need to be rented and the task schedule $\ell : c_i \to$ VM such that the minimum execution time of the analytical workflow is achieved:

$$\min_{all\ possibe\ \ell} (T_{total}) = \min_{all\ possibe\ \ell} \left(\sum_{all\ c_i \in CP} T(E_i) \right)$$

subject to the financial constraint:

$$C_{total} = \sum_{i=1}^{m} C_i \leq B$$

T_{total} and C_{total} denote the total execution time and the total monetary cost of a mapped analytical workflow, respectively, and C_i denotes the execution cost of component c_i. Both $T(E_i)$ *and* C_i are the function of n_i, see Eqs. (6) and (7) respectively. Since this work targets a single cloud, we do not consider the data transfer cost, i.e. $C_{tr} = 0$.

4   Adaptive Scheduling Algorithm

As most of tasks in analytical workflows are compute-intensive, of which the time of data transfer is much less than the execution time of the entire workflow, we assume that the data transfer time can be negligible.

The adaptive scheduling mechanism we proposed is performed in two stages: initial schedule for the analytical workflow based on heuristic strategy and VM adjustment stage.

4.1   Initial Schedule Algorithm

The pseudo code of initial schedule algorithm is provided in Algorithm 1. Starting with the least-tie schedule, we calculate the current critical path of the mapped workflow and only consider critical components for rescheduling. Because that each component may have multiple instances when executing, the execution time of the component is dynamical. Thus, different virtual machine numbers allocated for a component will resulted in different critical path for the overall workflow. The critical path calculation is repeatedly executed during the rescheduling process, and the number of iterations depends on the number of virtual machines and the budget.

We define the difference $\Delta Cost(c_i)$ and $\Delta T(c_i)$ in monetary cost and execution time, respectively.

$$\Delta Cost(c_i) = Cost_{cur} - Cost_{old} \qquad (8)$$

$$\Delta Cost(c_i) = T_{old} - T_{cur} \qquad (9)$$

The reschedule is allocated new number of virtual machine for the component with largest monetary cost difference $\Delta Cost(c_i)$ and minimal execution time difference $\Delta T(c_i)$. During the rescheduling process, the critical path may change, and in this case, we reschedule a new set of critical components and repeat the process until no rescheduling is feasible with the left budget.

Algorithm 1: *InitialSchedule* (G_c, K, T, B)

Input:

G_c: A workflow graph $G_c(V_c, E_c)$ with m nodes

K: Instance matrix of components

T: execution time matrix

B: Budget

Output:

N: VM numbers

ℓ_{time} : the time of optimal schedule

1: Find the *least-cost* schedule $\ell_{least-cost}$, which allocate one virtual machines to each component.

Calculate total cost as C_{min}, which is the lower bound cost, and

2: time T_{cmini}.

3: Find the *least-time* schedule $\ell_{least-time}$, which allocate $k_i \in K$ virtual machines to component c_i, and each component with minimal execution t_i.

4: Calculate total time as T_{min}, and total cost C_{tmin}.

5: **if** $B < C$ min

6:      **return** error ('no feasible solution for B').

7: **else**   min

8: **if** $C_{tmin} \le B$

9:      **return** $N = K$ and $\ell_{least-time}$.

10: **else**

11:      $C_{tmp} = C_{min}$.

12:      while $(C_{tmp} - B) > 0$

13:       Calculate the critical path under the current schedule.

14:       **foreach** component c_i in critical path.

15:           Decrease the allocated VMs until $cost_{c_i}^{new} < cost_{c_i}^{cur}$.

16:           Calculate the VM number difference Δn_i.

17:           Calculate the cost difference $\Delta C(E_i)$.

18:       **end**

19:       Sort $\Delta C(E_i)$.

20:       Reschedule the component with maximal $\Delta C(E_i)$.

21:           $C_{tmp} = C_{tmp} + \Delta C(E_i)$.

22:       **end**

23:      $N = K - \Delta n_i$.

24:      $\ell_{time} = EFT(c_m)$

25: **return** N and $\ell_{least-time}$.

26: **end**

4.2 VM Consolidation Algorithm

Provisioning and booting a new server with all required operating system packages and software applications will take some times. For example, it has been shown that AWS cloud servers require in average about 5 min to start a cloud server of small size (m1. small) and about 2 min to start high-CPU medium server (c1.medium) [20]. Thus, if we consolidate some virtual machines will reduce the star up and data transfer time. In our VM consolidation algorithm, we consolidate the virtual machines, in which are the critical components and has parent-child relationship (Algorithm 2). The main idea of the algorithm is very simple: for the instance task graph, we found out all of the sub-graphs and merge the virtual machines in the same sub-graph.

Algorithm 2: VMC $(G_c, N, Schedule)$

Input:

G_c: A workflow graph $G_c(V_c, E_c)$ with m nodes

N: VM numbers

$Schedule$: A workflow schedule

Output:

N_new: new VM numbers

$Schedule'$: new schedule

1:   $N_{new} = N$.
2:   Count=0;
3:   **get** *instance taks graph*.
4:   **Do sub-graph matching;**
5:   *Count=number of sub graphs.*
6:   **for all virtual machines**
7:     **The virtual machines in the same sub-graph will be consolidated.**
8:   **end**
9:   $N_{new} = N_{new} - Count$.
10:   **return** N_new **and** $Schedule'$.
11:

---

5 Case Study

We take KNN machine learning algorithm as example to illustrate the analytical workflow and evaluate the performance of our algorithm.

5.1 KNN-Based Classifier Workflow

In science, from bioinformatics to astrophysics to chemistry, the workflow for data mining algorithms are common. For illustration purposes, we use example of classification tasks, where a model is used to classify a set of test data. There are several widely-used approaches to building a model from a set of training data, such as decision trees (DT) and k-nearest neighbor (KNN). Within each approach several algorithms are possible. For example, decision tree algorithms include a classic divide and conquer algorithm (ID3) and a logistic model tree builder (LMT).

We will use KNN algorithm in our examples, which is a simple machine learning algorithm. Whenever we have a new point to classify, we find its K nearest neighbors from the training data. The choice of K will significantly influence the result of KNN algorithm. In general, the optimal K is given by scientist based on experience. We design a workflow model to automatic get the optimal K of KNN algorithm for a given dataset. Workflow model OA-KNN (Obtaining Accuracy of KNN) in Fig. 2 shows a dataflow structure where the maximum accuracy of KNN model is obtained for a set of test data [18]. This workflow model consists of three sequence components, which are data processing, parameter tuning and KNN classifier. The data processing component is responsible for loading dataset (read digitals and labels) and formatting the data. Parameter tuning component is to find the optimal k of the KNN model and the task of KNN classifier component is classification. When interpreted by workflow engine, the parameter component will generate multiple instances. The execution time of each component depends on the dataset, and the analysis task usually is time consuming in bid data area.

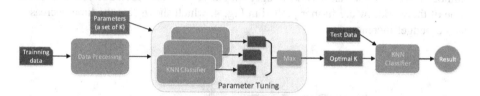

Fig. 2. KNN-based classifier workflow

5.2 Simulation Setting

The dataset is MINIST, which was developed by Yann LeCun, Corinna Cortes and Christopher Burges for evaluating machine learning models on the handwritten digit classification problem [19]. The dataset was constructed from a number of scanned document dataset available from the National Institute of Standards and Technology (NIST). Each image is a 28 by 28 pixel square (784 pixels total). A standard spit of the dataset is used to evaluate and compare models, where 60,000 images are used to train a model and a separate set of 10,000 images are used to test it. It is a digit recognition task. As such there are 10 classes (10 digits) to predict. Results are reported using prediction error, which is nothing more than the inverted classification accuracy.

By using the KNN-based classifier workflow, the optimal K can be found and 10000 images will be recognized with high prediction accuracy. We statics that the average execution time of three components, see Table 1. The price of the virtual machines is set to $3.837 per hour and the average start up and cool down time is set to 5 min and 3 min, respectively.

Table 1. Average execution time of digit classification workflow

Component	Instance number	Average processing time (hour)
Data processing	1	0.113
Parameter tuning	k	[1.027, $k*1.027$]
KNN classification	1	0.418

5.3   Simulation Results

First of all, for illustration purposes, we consider a number of parameter $k = 12$ and a budget equals to 70 dollars (Fig. 3). Then we compute the schedule results with *InitialSchedule* (Algorithm 1) and with VM consolidation algorithm, respectively. Figure 3 shows the schedule time of the virtual machines. The rectangular bars represent the renting time of the virtual machines and slashes areas of the bars denote the factorial utilizing time. Clearly, the schedule with virtual machine consolidation can reduce the number of the rented virtual machines and improve the virtual machine utilization.

Give the value of the parameter, we can get the valid range of the budget. We list in Table 2 all schedules found by the algorithm as the budget varies within all real numbers. We can get that a valid budget should bigger than \$57.555. The execution time of the workflow are further plotted in Fig. 4, which shows that the time decreases as the budget increase.

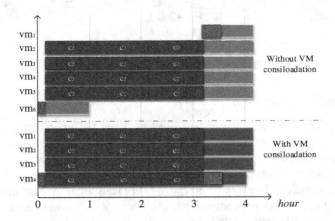

Fig. 3. VM schedule time ($k = 12$, budget = \$70)

In Figs. 5 and 6, we observe the relationship between monetary cost (workflow execution time) and the number of virtual machines. Obviously, the more virtual machines the less execution time. However, the monetary cost is not monotonic increase with the number of virtual machines because of the pricing mechanism partial instance-hour consumed will be billed as a full hour.

For a more comprehensive analysis of the algorithm, we observe the schedule results with different budgets and parameter values.

Table 2. Schedule result with different Budgets

Budget	InitalSchedule		With VM consolidation	
	VMs	Time (h)	VMs	Time (h)
$(\infty, 57.555)$	null	null	null	null
$[57.555, 61.392)$	3	12.775	1	12.549
$[61.392, 65.229)$	4	6.613	2	6.387
$[65.229, 69.066)$	5	4.559	3	4.333
$[69.066, 76.740)$	6	3.532	4	3.306
$[76.740, 99.762)$	8	2.505	6	2.279
$[99.762, \infty)$	14	1.478	12	1.252

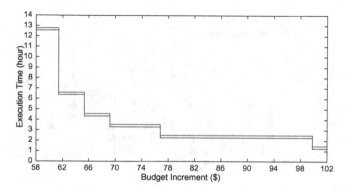

Fig. 4. Time measurements with different budgets ($k = 12$)

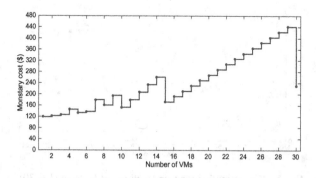

Fig. 5. The cost of parameter tuning component under different number of VMs ($k = 30$)

Figures 7 and 8 show that the allocated number of virtual machines and the execution time under different budgets, respectively with parameter value k = 30. Clearly, the more budget, the more virtual machines can be rented as well as lower execution time.

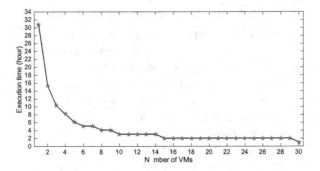

Fig. 6. The time of parameter tuning component under different number of VMs ($k = 30$)

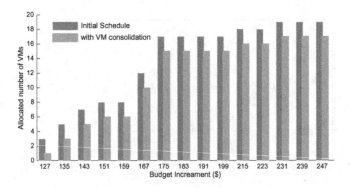

Fig. 7. The allocated VM numbers under different budgets ($k = 30$)

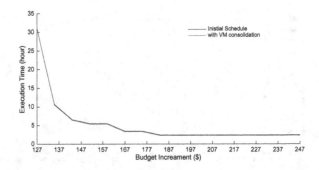

Fig. 8. The execution time under different budgets ($k = 30$)

And the algorithm with virtual machine consolidation has better result than only initial scheduling.

Next, we change the parameter values. The left plots are the results that changing budgets according to the lower-bound of the budgets with individual k. And the right

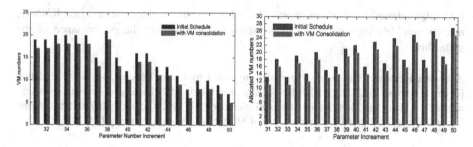

Fig. 9. VM numbers with different number of parameters

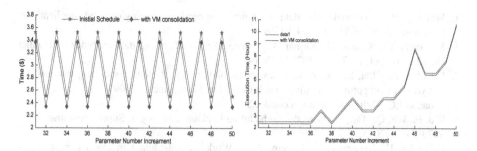

Fig. 10. Execution time with different number of parameters

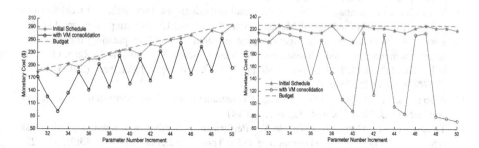

Fig. 11. Monetary cost with different number of parameters

plots are the results with budget equals to $226.5. We can see that for all results, the algorithm with VM consolidation algorithm is better (Figs. 9, 10 and 11).

6   Conclusion

In this paper, we investigated how to allocate virtual machines for analytical work flows, with the goal of minimizing the execution time as well as meeting budget constraints at the same time. We formulated the budget-constrained scheduling problem for analytical workflow. We proposed a two stages heuristic algorithm and take KNN (K nearest neighbor algorithm) as a case study to demonstrate the performance of the algorithm.

Acknowledgement. This work is partially supported by China National Science Foundation (Granted Number 61272438, 61472253), Research Funds of Science and Technology Commission of Shanghai Municipality (Granted Number 15411952502, 14511107702) and Cross Research Fund of Biomedical Engineering of Shanghai Jiaotong University (YG2015MS61).

References

1. Watson, H.J.: Tutorial: big data analytics: concepts, technologies, and applications. Communications of The Ais 34.1 (2014)
2. Kambatla, K., Kollias, G., Kumar, V., Grama, A.: Trends in big data analytics. J. Parallel Distrib. Comput. **74**(7), 2561–2573 (2014)
3. Rimal, B.P., Choi, E., Lumb, I.: A taxonomy and survey of cloud computing systems. In: Networked Computing and Advanced Information Management (2009)
4. Amazon EC2. http://aws.amazon.com/ec2/
5. Wu, F., Wu, Q., Tan, Y.: Workflow scheduling in cloud: a survey. J. Supercomputing **71**(9), 3373–3418 (2015)
6. Fakhfakh, F., Kacem, H.H., Kacem, A.H.: Workflow scheduling in cloud computing: a survey. In: Enterprise Distributed Object Computing (2014)
7. Singh, L., Singh, S.: A survey of workflow scheduling algorithms and research issues. Int. J. Comput. Appl. **74**(15), 21–28 (2013)
8. Yu, J., Ramamohanarao, K., Buyya, R.: Deadline/budget-based scheduling of workflows on utility grids. In: Proceedings of Market-Oriented Grid and Utility Computing, pp. 427–450
9. Zheng, W., Sakellariou, R.: Budget-deadline constrained workflow planning for admission control in market-oriented environments. In: Vanmechelen, K., Altmann, J., Rana, Omer, F. (eds.) GECON 2011. LNCS, vol. 7150, pp. 105–119. Springer, Heidelberg (2012). doi:10. 1007/978-3-642-28675-9_8
10. Arabnejad, H., Barbosa, J.G.: A budget constrained scheduling algorithm for workflow applications. J. Grid Comput. **12**, 1–15 (2014)
11. Topcuoglu, H., Hariri, S., Wu, M.Y.: Performance-effective and low-complexity task scheduling for heterogeneous computing. IEEE Trans. Parallel Distrib. Syst. **13**(3), 260–274 (2002)
12. Lin, X., Wu, C.Q.: On scientific workflow scheduling in clouds under budget constraint. In: Proceedings of 42nd International Conference in Parallel Processing (ICPP), pp. 90–99. IEEE (2013)
13. Sakellariou, R., Zhao, H., Tsiakkouri, E., Dikaiakos, M.D.: Scheduling workflows with budget constraints. Proceedings of Integrated Research in GRID Computing, pp. 189–202. Springer, New York (2007)
14. Zeng, L., Veeravalli, B., Li, X.: Scalestar: Budget conscious scheduling precedence-constrained many-task workflow applications in cloud. In: Proceedings of IEEE 26th International Conference on Advanced Information Networking and Applications (AINA), pp. 534–541. IEEE (2012)
15. Amazon AWS. http://aws.amazon.com/ec2/instance-types/. Accessed 25 April 2015

16. Microsoft Azure. http://azure.microsoft.com/en-us/services/virtual-machines/. Accessed 25 April 2015
17. Google Compute Engine. https://cloud.google.com/compute/. Accessed 25 April 2015
18. Gil, Y., Groth, P., Ratnakar, V., Fritz, C.: Expressive reusable workflow templates. In: Proceedings of Fifth IEEE International Conference on e-Science, pp. 344–351 (2009)
19. LeCun, Y.: The MNIST database of handwritten digits. http://yann.lecun.com/exdb/mnist/index.html
20. Suleiman, B., Sakr, S., Venugopal, S., Sadiq, W.: Trade-off analysis of elasticity approaches for cloud-based business applications. In: Wang, X.S., Cruz, I., Delis, A., Huang, G. (eds.) WISE 2012. LNCS, vol. 7651, pp. 468–482. Springer, Heidelberg (2012). doi:10.1007/978-3-642-35063-4_34

Data Driven Service Computing

Monitoring as a Service Based on Pub/Sub System over a Cloud Environment

Dingyu Yang[✉] and Chunlei Ji

Shanghai Dian Ji University, Shanghai, China
{yangdy,jicl}@sdju.edu.cn

Abstract. A cloud environment always includes multiple cloud providers, which have their own resources and can access each other with specified methods. However, it is challenging to monitor an application deployed in such a environment due to their heterogeneous metadata and interfaces. In this paper, we develop a pub/sub monitoring system over a cloud environment. Several probes are designed to collect data and we apply complex event processing to reduce network traffic. We propose an adaptive method to adjust the monitoring frequency, which can save communication cost and not loss the most feature of monitoring data. Moreover, we develop a new method NBS to balance the workload of our cluster and it can improve the performance of our system. We deploy our system in three clouds and the results show that our methods can improve the throughput of the monitoring system.

Keywords: Distributed monitoring · Pub/Sub system · Cloud computing · Adaptive frequency

1 Introduction

The emergence of cloud computing, which offers computing infrastructure, platform, and applications as services [1,5] to one or more tenant organizations. They provide infrastructure service to tenants by virtualized resource. The design model eliminates the upfront capital costs and in-house operation costs for the tenants. In order to abstract more and more tenants, cloud providers buy lots of infrastructure to ensure the sufficient resources. To manage these resources effectively, a monitoring system which can collect and report on the behavior of these resources is needed.

There are several available cloud providers, such as Amazon Elastic Compute Cloud [5] and Aliyun [1]. However, different cloud solutions are rarely compatible with each other and this creates a kind of vendor lock-in, which is not only limiting to the customer, but also limits the potential of cloud as a whole. A federated cloud [6] is one where competing infrastructure providers can reach cross-site agreements of cooperation regarding the deployment of service components in a way similar to how electrical power provides provision capacity from each other to cope with variations in demand. As [19] says, a federated

© Springer Nature Singapore Pte Ltd. 2017
J. Cao and J. Liu (Eds.): MiPAC 2016, CCIS 686, pp. 49–64, 2017.
DOI: 10.1007/978-981-10-3996-6_4

cloud should be technological capabilities to federate disparate data centers, including those owned by separate organizations. Only through federation and interoperability can infrastructure providers take advantage of their aggregated capabilities to provide a seemingly infinite service computing utility [6].

The federation gives monitoring new challenges for its heterogeneous metadata and interfaces. First one is that it is difficult to control and manage the cloud resource. The design paradigm (Iaas, PaaS, SaaS) [2,9] separates hardware provider and software provider. Software is run by the SaaS provider, and customers do have less control over the application. Resource management is also limited by IaaS providers. Secondly, there is no choice to monitor some key resources or to subscribe a service personally. The given monitoring information is insufficient and cannot fulfill user's requirement. Thirdly, the uniform monitoring objects and schemas are static and difficult to be extended. For example, the monitor frequency is an importance parameter and always fixed in current cloud platforms [1,5].

We adopt a pub/sub monitoring system in a cloud computing environment. Pub/Sub system is widely used in distributed system and distributed information dissemination [3,8]. It has high scalability and elastic to quickly adapt to workload changes. Pub/sub mechanism makes the monitoring more flexible as follows:

When an application is deployed in heterogeneous cloud environment, the monitoring is more critical for the various interfaces. The federation of physical and virtualized resources is making the monitoring more complex. A pub/sub middleware can integrate different data sources and interfaces to generate a uniform data schema. For example, a service may rely on hardware and software distributed over many sites, we can aggregate monitoring data from these sites on demand. The middleware is convenient to adapt the change of certain cloud and scalable for the new customers or providers. Pub/Sub system can provide an asynchronous communication pattern. It can be automatically notified when new data becomes available. It can filter some unnecessary data or transform the status of the data. Compared to a traditional centralized client/server communication model, pub/sub can reduce the network traffic and computational overhead. Data events are published and disseminated to interested recipients. In a cloud data center, the network delay is small and the packet loss rate is low. When loss of Internet connectivity or data is inaccessible, the middleware can create a temporal storage of the sub-data.

Therefore, this paper designs a pub/sub system to monitoring cloud environment. It provides a monitoring service on demand. Pub/sub system is a processing and transmitting layer. Sensors are deployed in cluster or federated clouds, which are responsible for collecting and publishing data to the network. The monitoring data are wrapped into events with some attributes. Complex event processing (CEP) engine are also designed to filter or aggregate data events. Users submit their requirements as subscriptions to pub/sub system. The system partitions the subscriptions in multiple segments and deploys them in a cluster. When data events are delivered in the network, pub/sub system firstly

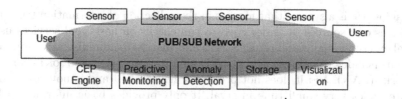

Fig. 1. The structure of Pub/Sub network

matches the subscriptions and then diverts the events to interested subscribers. The following Fig. 1 is a simple structure of Pub/Sub network.

In summary, the contributions of this paper are as follows:

1. We develop a pub/sub monitoring system to monitor the status of cloud environment. It provides an asynchronous mechanism to transfer events. Complex event processing is adopted to filters and aggregates atom events, which reduces network events and improves the performance of our system.
2. We propose an adaptive method to adjust the monitoring frequency. The adjustment is based on data-driven model and decided by the data distribution. The frequency is auto-changed if current configure is not suitable. The adaptive frequency can eliminate the network traffic and not loss the most feature of monitoring data.
3. We design a neighbor-balance strategy to adjust the deployment of subscription. The strategy modifies the subscriptions based on the statistic of throughput and runtime. It is handled automatically without manual configuration. The loads of overloaded nodes are alleviated and less-loaded ones are given more tasks. It can balance the load of processing nodes and improve the throughput of our system.
4. We develop the monitoring system, and deploy it in three clouds. Extensive experiments have been conducted to show the efficiency of the system. Our results show that Balance method NBS performs better than existing method *Random* method.

This paper is organized as follows. Existing related work is reviewed in Sect. 2. In Sect. 3, we design the pub/sub monitoring system. The system improvement is presented in details in Sect. 4. Extensive experiment results are reported in Sect. 5. Finally, in Sect. 6, we conclude the whole paper.

2   Related Work

Ganglia [14] uses a hierarchical system where the attributes are replicated within clusters using multicast and then cluster aggregates are further aggregated along a single tree. It is based on a design targeted at federations of clusters. The system has widely technologies and algorithms to achieve low overhead. There are some aspects needed to be regarded, such as load balance.

Nagios [15] is a system that focuses in the collection of information regarding the status of resources. It is designed to run checks on hosts and services using several external plugins and return status information to administrative contacts. While it includes valuable features and capabilities, it is destined for Local Area Networks (LAN), and it does not provide a generic API and is not designed to operate under very small.time interval. It only provides basic information but cannot support users exact requirements.

Clayman et al. [4] create a monitoring framework Lattice for service clouds which succeeds in the scalability of the monitoring. The framework is spread in different layers (service, virtual environment, physical resources, etc.). It offers probes to collect data on physical infrastructures, virtual resources and service applications. The federation of VEEs (virtual execution environments) for multiple services is also presented. However, the probe is not considered the rate of collection. And they do not provide the details in processing the data and network transfer.

Monalytics [10] is an online monitoring system integrated monitoring and analytics to effectively manage large scale data centers. The system has multiple levels to monitoring target systems. Data Capture Agents collect data and Brokers aggregate and analysis outputs from multiple agents. Zone level is a collection of nodes associated with the brokers on each physical node. It proposes an interesting architecture and many novel ideas. Nevertheless, the system does not take the users demand into account because different people need different data.

Chukwa [18] was built on top of Hadoop and proposed log collection framework and network management system. It stored the log data using HDFS. The collection tasks are executed from a larger number of agents. A MapReduce job periodically compacts and merges the files. On the other hand, the system is based on collecting the log generated by Hadoop system. The federated cluster is unconsidered and there not exist a global cloud log system.

Amazon Simple Notification Service (SNS [5]) is a web service that offers a pub/sub model to meet the customers needs. It only provides an API based on Amazon cloud and is hard to spread to other data centers. Facebook Scribe [20] is a server for aggregating streaming log data. It is designed to scale to a very large number of nodes and be robust to network and node failures. There is a scribe server running on every node in the system, configured to aggregate messages and send them to a central scribe server (or servers) in larger groups. However, the system is based on a central server to send messages. If the server is not available, the message would be lost. And the central server has high throughput and busy traffic.

Previous efforts focus on data collection and aggregation. The probes are predefined and deployed as a rigid schema. The data is transferred to users in a uniform template. The current system is difficult to extend when users want different data or some other interesting data. Moreover, collecting data frequency is often solid or periodical. This is not proper if the data sequence is stable and some data is repeatedly collected. The dispatch task is always executed in a centralize server. It leads to high network traffic and overhead. If the server is broken down, the monitoring messages would be lost.

3   System Design

In order to provide a scalable, elastic, autonomic and federate system, we design the monitoring system with multiple components: *Probe and Event, Pub/Sub System, Subscriptions Storage, Monitoring Service* and *Application*. The publish/subscribe framework shown in Fig. 2 is motivated by BlueDove [11], which is a scalable and elastic publish/subscribe service.

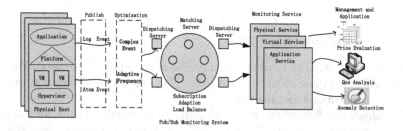

Fig. 2. The framework of Pub/Sub monitoring system

3.1   Probe and Event

Probes are designed and deployed in the cloud and data center. The probes capture the states of the target system (e.g., OS, virtual machine, cloud platform and application) and collect required data. Some probes also collect data information from log files in hadoop cluster. The data includes the executing behaviors of hadoop (e.g., task execution time, job waiting time and available disk space). The event is very simple and is the atom element of our system. It is wrapped in desired format, which is convenient to filter some unnecessary data and aggregate data. For example, if we are interested in the CPU load data with more than 80, some data such as $CPU < 80$ would be discarded and not stored in our system.

An event consists of header, attributes and open contents. As we see in Fig. 3, each element has specific definition and information. The header consists of meta-information about the event, such as the occurrence time. The attribute part contains specific information about the occurrence itself (e.g. CPU load/float/common attribute). An event can also contain free-format open content information. An event definition may contain references to other events when there is semantic relationship between them. There are four types of relationships: membership, generalization, specialization and retraction [7].

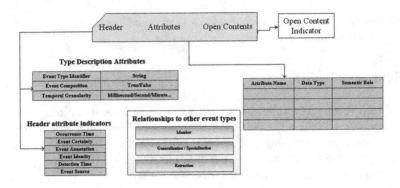

Fig. 3. The format of event element

3.2   Pub/Sub Monitoring System

Our system monitors the resource from public cloud and private cloud. We have written probes for these monitoring objects. There are physical resources, virtual resources, applications and log files.

The physical information in a host includes basic information and dynamical resources. Basic information has fixed value and generally do not changes frequently, such as host name, IP, operation system, CPU type and capacity, memory type and capacity, disk type and capacity, network bandwidth, and so on. The probes monitoring basic information collect data periodically. Dynamical resources are always changing and hard to be forecasted. The probes regarding dynamical resources needs to collect data frequently or in real time. Dynamical resources includes CPU usage, memory usage, free memory, disk usage, network usage.

Virtual resources are similar to physical monitoring. The probes get the data not from interior virtual system, but interact with VM hypervisor. The probes collect data (e.g., CPU usage, memory usage, free memory, disk usage, network usage) via the interface of VM hypervisor.

Since all services or applications are executed in virtual machines, we deploy our probes running in virtual machines. We publish some probes to collect data from log files. Logs provide a glimpse into the states of a running system. For example, the runtime of tasks and jobs in hadoop can be extracted from the logs. Our probes can monitor the logs in federated clouds and get the interested data.

Customers always deploy their application on federated clouds. For example, in order to get some stock data in other countries in real-time, financial developers create a Amazon EC2 instance and publish their program to crawler data. Meanwhile, they use a domestic cloud service for their analysis application. At this time, if customers want to know the status of all instances in the federated clouds. Our probes can be deployed crossing cloud to monitor the resources and send data events to pub/sub system. The communication mechanism provides an asynchronous paradigm to improve the efficiency.

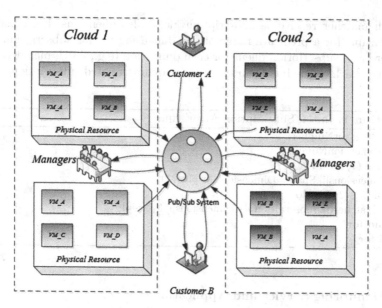

Fig. 4. An example of monitoring federated cloud

Figure 4 describes an example of monitoring federated cloud. There are two cloud providers (Cloud 1 and Cloud 2). Every host has multiple virtual machines, which are established according to customers requirements. Federated clouds can efficiently meet load peaks for limited resources. Pub/Sub monitoring system can provide an interested point of view of detecting the changes or anomaly. Customers deploy services to another collaborator cloud when the resources are insufficient. The managers of a cloud provider subscribe their interested data to estimate the resource utilization.

3.3 Subscriptions Storage

In order to provide high output, the system uses a multi-dimensional subscription space partitioning approach to organize subscriptions. Each dimension is divided into many continuous subset spaces. Each subset space is assigned to one matcher, which is responsible for searching in this space. For example, CPU load is between 0 and 100 and there are five matchers in the network. Then in the initial process program, CPU subscription is divided into five ranges (e.g., $[0, 20), [20, 40), [40, 60), [60, 80), [80, 100))$. The detail of initialing subscriptions is shown in Algorithm 1.

Given a subscription $\mathbb{S} = S^1 \times \cdots S^k$, a dispatcher assigns $\mathbb{S}$ to matchers k times, each time along a different dimension. When the matcher is suitable for the subscription, the copy of the subscription $\mathbb{S}$ would be sent to the matcher. Each subscription has at least copies stored on different matchers, which provides a natural means of fault-tolerance.

Each matcher receives a subscription along all dimensions. Each matcher is responsible for a predicate range of one dimension in this subscription. The matcher has a subscription queue for each dimension (e.g., CPU, memory, Task Runtime). Each queue builds an index for searching related subscriptions.

Algorithm 1. Initial Subscription Algorithm

1: M= MatcherNumber;
2: D=DimensionNumber;
3: **for** $i \in (1, D)$ **do**
4:    Dimensions[i].Space=Dimensions[i].split(M);
5:    **for** $j \in (1, M)$ **do**
6:       Matcher[j].Subscription[i]=Dimensions[i].Space[j];
7:    **end for**
8: **end for**

3.4 Monitoring Service and Application

Our system provides multiple metrics, such as physical metrics (e.g., CPU, memory, I/O, and network bandwidth), virtual metrics (e.g., virtual machines, virtual CPU, memory, storage, network utilization), and application metrics (e.g., OS, hypervisor, database, workload, service, task runtime, job runtime). The resources in the cloud environment are diversity. Cloud users subscribe their interested metrics and submit the subscriptions to the pub/sub system. The system provides a service to integrate the subscriptions and sends the matching result to the user automatically.

From the monitoring service, we can provide interested data, such as the status of hardware and software, to users to estimate their infrastructure. Furthermore, we can give some advice to cloud managers for their serving quality. For example, some statistic of resource usage can be analyzed to determine the cloud price. Quality of service is a factor to evaluate a system. The runtime, buffer size, communication policies and loss rate are related to quality of service and should be monitored in real-time.

4 System Improvement

4.1 Complex Event Processing

In real environment there are primitive events transferred to the network, especially in a heterogeneous environment. We use primitive events to integrate various monitoring data from multiple cloud providers. It is useful for heterogeneous data integration. From the history monitoring data, we find that there exists some events having similar header or metadata. In order to maximize the network throughput, we introduce complex event processing engine (CEP) to improve the performance.

Complex Event Processing is a defined set of tools and techniques for analyzing and controlling the complex series of interrelated events [13]. There are several functions of CEP (e.g., Filtering, Transformation). Filtering is selecting which of the input events participate in the processing via the logical condition. Matching is also done to find interested events using some kind of matching criterions. Transformation is taking input events and creating output events that are function of these input events. Transformation has several different types, such as Translate, Enrich, Project, Aggregate, Split and Compose.

Figure 5 shows the detail of these operations [13]. *Translate* takes a single event as its input, and generates a single derived event which is a function of the input event. *Enrich* creates a derived event which includes the attributes from the original event, possibly with modified values, and can include additional attributes. *Project* creates a single derived event containing a subset of the attributes of the input event. *Aggregate* takes as input a collection of events and creates a single derived event by applying an aggregation function over the input events. *Split* performs a single event in, multiple events out. *Compose* takes groups of events and creates derived events using a criterion.

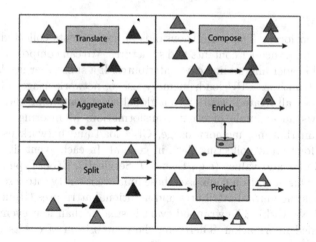

Fig. 5. Event transformation

In traditional monitoring system, it is difficult to get some high level information on demand (e.g., aggregated data). It is always done offline and sent to interested users. CEP provides a convenient method to online aggregate data. For example, when a user wants to know the average CPU load or temperature of a certain computer room, CEP collects data events of each host in this room and calculates the average value. The result will be sent to the interested user by pub/sub network.

In federated cloud environment, applications are always deployed in several providers. Monitoring services have to integrate the heterogeneous data from

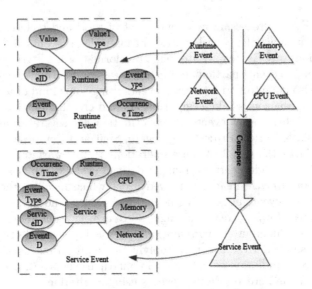

Fig. 6. An example of event compose service

other cloud platforms. There are thousands of atom events collected and some events have similar header contents and structures. We can compose these events using mutual header and adding extra attributes. If a subscriber needs the monitoring data (runtime, CPU load, memory usage, network usage) of a service, we can compose all related events into a global event containing these contents. Figure 6 shows an example of event transformation: Monitoring metrics of a service contain runtime, memory usage, CPU load and network usage. These events need four event collecting data in general. In each event, it has several similar attributes and values (e.g., EventID, ServiceID, EventType and Occurrence Time). Our operation composes these four event entity into a service event which contains the mutual attributes and additional attributes (Runtime, CPU, Memory and Network). The wrapped event is smaller than four events but contains all needed information. Therefore, the composition of events can reduce some network communication cost.

4.2 Adaptive Monitoring Frequency

A monitoring system has its own cost, such as data generation, collection, store or processing cost. It should be low overhead, and not affect the performance of other applications. Otherwise, it will lose its monitoring value. Monitoring frequency is an important aspect that impact the tracing overhead. The monitoring with high frequency can provide an efficient schema to capture the interested data, but it has the potential to greatly reduce the workload on the servers [20]. However, lower traffic workloads may miss important events and cannot trace the status of cloud environment. How to determine the monitoring frequency is challenging.

We propose an adaptive solution to adjust the frequency. It modifies the frequency according the distribution of the data. That means the model is dynamically adjusted by identifying whether the collection rate is proper. When the rate is lower than expectation, the frequency would be increased. Otherwise, the rate would be cut down.

We provide an error range $[Min\_Error, Max\_Error]$ to estimate the collection rate in a time window. We calculate the error by mean relative error (MRE). If the average error is more than Max_Error, the frequency is auto-adjusted to increase the value. Otherwise, if the average error is less than Min_Error, the frequency will be reduced. Support n is the data number in a time windows and we can estimate the MRE by Formula 1. The details of our method is described in Algorithm 2.

$$MRE = \frac{1}{n} \sum_{i=1}^{n} \frac{|\hat{x}_i - x_i|}{x_i} \tag{1}$$

Algorithm 2. Adaptive Frequency Algorithm

1: Rate= OriginalValue;
2: D=IncrementValue;
3: Max_Error=max;
4: Min_Error=min;
5: **while** $waitinganinterval$ **do**
6:     OriginalValues=CollectSequence();
7:     EstimationValues=EstimatedSequence();
8:     n=OriginalValues.length
9:     $mre = \frac{1}{n} \sum_{i=1}^{n} \frac{|EstimationValues_i - OriginalValues_i|}{OriginalValues_i}$
10:     **if** $mre > Max\_Error$ **then**
11:         Rate=Rate+D;
12:     **else if** $mre < Min\_Error$ **then**
13:         Rate=Rate-D;
14:     **end if**
15: **end while**

4.3   Subscription Load Balance

The data distribution in real world is often skewed. The Pareto principle states that, for many events, roughly 80% of the effects come from 20% of the causes [17]. CPU load also follows this rule. It always centralizes a certain range. For example, we analyze the data from CMUs clusters collected by Dinda [12]. More than 80% CPU loads in host $apx0$ and $apx7$ are between 0 and 4%. Since our subscriptions are deployed by the range of dimensions, it makes the nodes containing the subscriptions between 0 and 4% very busy and other nodes are idle.

It also leads to the matching tasks very slow in busy nodes and degrades the performance of the whole system.

We present a neighbor-balance strategy (NBS) to adaptively adjust the segments of dimensions. The idea of this approach is that two adjacent hosts exchange the subscriptions if they have greater load differences. For example, host As load is 20% more or less than its neighbor Bs load, then we can balance them using exchanging their subscriptions. The target is to move some subscriptions from overloaded nodes to less-loaded ones and balance the overload of all the matching servers. The dispatchers record the numbers of dispatching messages in every matching server. NBS is adaptive adjusting the distribution of subscriptions. When detecting the great difference between a host and its neighbor, it would reassign the segments by analyzing the distribution of dimensions. The segments are calculated by averaging the overload of both hosts.

The adjacent-segment hosts exchanging information have some advantages. The neighbor segments can be split or merged facilely. The continuous subscriptions can be merged into an entire subscription. For example, an original CPU subscription ($15 < CPU < 25$) has been divided into two subscriptions ($15 < CPU < 20$, $20 < CPU < 25$) because the CPU subscription is divided into five equal segments: $[0, 20), [20, 40), [40, 60), [60, 80), [80, 100)$. After the load of host ($[0, 20)$) is much more than that of host $[20, 40)$, NBS will be triggered to split or combine subscriptions. After combining the segments: $[0, 10), [10, 40),$ $[30, 60), [60, 80), [80, 100)$, the subscription can be merged as its original form ($15 < CPU < 25$). Split and merge subscriptions between neighbor segments makes relative segments join together. It reduces transferring subscriptions and makes the operation effectively. The detail of the method is shown in Algorithm 3.

5   Experiment Evaluation

We have implemented the monitoring system and deployed our probes in multiple cloud providers, such as Aliyun [1], NewTouch Cloud [16] and our private cluster. Our private cluster is applied for our private files and data. Our web client is deployed in Aliyun, which has enough network bandwidth to support the workload. Hadoop system is deployed in NewTouch Cloud and our analysis system is also in this cloud.

We developed some probes, such as CPU Load, Memory Usage, Disk Usage, operation system, database and logs of Hadoop. The probes can be remotely configured in back-end machines. We can dynamically modify the configuration and control the probes on demand.

5.1   Adaptive Monitoring Frequency

Since the monitoring task has its own overhead, how to reduce the overhead is an issue of current monitoring system. As we know, the frequency is the rate of collecting and transmitting monitoring data. It affects the number of network messages and performance of our system. If the frequency is controlled in a

Algorithm 3. NBS Algorithm

```
1: Input:Host A, Host B
2: HighLoad=A.load;
3: LowLoad=B.load;
4: HighStartIndex=A.StartIndex;
5: HighEndIndex=A.EndIndex;
6: LowStartIndex=B.StartIndex;
7: LowEndIndex=B.EndIndex;
8: if HighStartIndex = LowEndIndex then
9:    HighLength=(HighEndIndex-HighStartIndex)*(HighLoad+LowLoad)/(2* HighLoad);
10:   HighStartIndex= HighEndIndex- HighLength;
11:   LowEndIndex= HighStartIndex;
12: else if HighEndIndex = LowStartIndex then
13:   HighLength=(HighEndIndex- HighStartIndex)* (HighLoad+LowLoad)/(2* HighLoad);
14:   HighEndIndex= HighStartIndex+ HighLength;
15:   LowStartIndex= HighEndIndex;
16: end if
17: A.StartIndex=HighStartIndex;
18: A.EndIndex=HighEndIndex;
19: B.StartIndex=LowStartIndex;
20: B.EndIndex=LowEndIndex;
```

reasonable rate, then we can fulfill our requirement of subscriptions and reduce the overhead as possible.

Every probe has a frequency attribute and it has a default value (e.g., 1 s). It means every second, the probe collects the data and wraps the data into an event with some information. In order to dynamic adjust the frequency, we set an estimated range $[Min\_Error, Max\_Error]$, which denotes that if the estimated error is less than Min_Error, or more that Max_Error, the frequency will be readjusted. In our experiment, we set the range as $[0.5\%, 5\%]$. We monitor the two metrics: $CpuLoad$ and $DiskUsage$. We can find that in Fig. 7, from the original data of one node, cpu load is more fluctuant than disk usage.

After we adopt the adaptive monitoring frequency mechanism, the probe can self-adjust the frequency based on the distribution. We calculate the message number and compare with the original data sequence. We find that our method can reduce the message number in a reasonable error range. The result in Fig. 8 shows that $CpuLoad$ can save 26% communication cost and $DiskUsage$ is more efficient and can save 53% cost. That is because $DiskUsage$ is more stable that $CpuLoad$ and the frequency is smaller.

5.2   Subscription Load Balance

The matching task in pub/sub monitoring system is based on the subscriptions. That means when one node receive a event, it iterates the subscription list to find the subscribers. The overhead of this node is also sensitive with the number of the subscriptions. If the number is very large, the task matches very slowly.

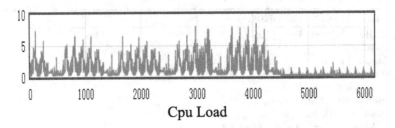

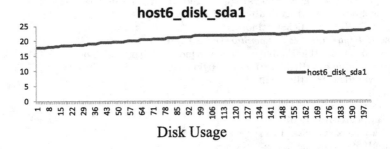

Fig. 7. Cpu load and disk usage of one node

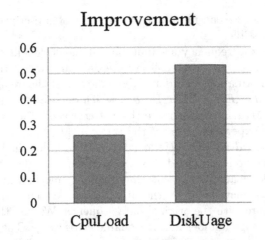

Fig. 8. Improvement of adaptive monitoring frequency

In our experiment, we randomly generate 10000 subscriptions and deploy them in five nodes. Each of them are initially located based on the default dimension range. For example, for Cpu Load, the range is splinted into five equal ranges. After a while, we find that the cpu load in five nodes are different, and node 2 is overloaded, and node 3 is relative idle. Our method will be triggered to balance the range based on the data and the result shows that the load of five nodes are more uniform in Fig. 9.

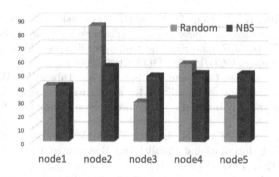

Fig. 9. The result of NBS

6 Conclusion

In this paper, we studied the problem of monitoring a cloud environment and implemented the system based on pub/sub mechanism. We designed multiple probes and deployed them in three clouds. Two mechanisms are proposed to improve the performance of our system. Extensive results show the efficiency of the system. Specifically, our methods can reduce 20%−50% network traffic and balance the workload of our processing nodes by dynamically exchanging the subscriptions.

Acknowledgment. Supported by Academic Discipline Project of Shanghai Dianji University, Project Number: 16YSXK04.

References

1. Aliyun (2012). https://www.aliyun.com/
2. Armbrust, M., Fox, A., Griffith, R., Joseph, A.D., Katz, R., Konwinski, A., Lee, G., Patterson, D., Rabkin, A., Stoica, I., et al.: A view of cloud computing. Commun. ACM **53**(4), 50–58 (2010)
3. Baldoni, R., Virgillito, A.: Distributed event routing in publish/subscribe communication systems: a survey. DIS, Universita di Roma La Sapienza. Technical report, 5 (2005)
4. Clayman, S., Galis, A., Chapman, C., Toffetti, G., Rodero-Merino, L., Vaquero, L.M., Nagin, K., Rochwerger, B.: Monitoring service clouds in the future internet. In: Future Internet Assembly, pp. 115–126 (2010)
5. Amazon EC2 (2010). http://aws.amazon.com/ec2
6. Elmroth, E., Larsson, L.: Interfaces for placement, migration, and monitoring of virtual machines in federated clouds. In: Eighth International Conference on Grid and Cooperative Computing, pp. 253–260. IEEE (2009)
7. Etzion, O., Niblett, P.: Event Processing in Action. Manning Publications Co., Greenwich (2010)
8. Eugster, P.T., Felber, P.A., Guerraoui, R., Kermarrec, A.-M.: The many faces of publish/subscribe. ACM Comput. Surv. (CSUR) **35**(2), 114–131 (2003)

9. Fox, A., Griffith, R., Joseph, A., Katz, R., Konwinski, A., Lee, G., Patterson, D., Rabkin, A., Stoica, I.: Above the clouds: a berkeley view of cloud computing. Department Electrical Engineering and Computer Sciences, University of California, Berkeley, Report UCB/EECS, 28(13):2009 (2009)

10. Kutare, M., Eisenhauer, G., Wang, C., Schwan, K., Talwar, V., Wolf, M.: Monalytics: online monitoring and analytics for managing large scale data centers. In: Proceedings of the 7th International Conference on Autonomic Computing, pp. 141–150. ACM (2010)

11. Li, M., Ye, F., Kim, M., Chen, H., Lei, H.: A scalable and elastic publish/subscribe service. In: IEEE International Parallel and Distributed Processing Symposium (IPDPS), pp. 1254–1265. IEEE (2011)

12. C. Load (2003). http://www.cs.northwestern.edu/pdinda/loadtraces/

13. Luckham, D.: The power of events: an introduction to complex event processing in distributed enterprise systems. In: Bassiliades, N., Governatori, G., Paschke, A. (eds.) RuleML 2008. LNCS, vol. 5321, p. 3. Springer, Heidelberg (2008). doi:10.1007/978-3-540-88808-6_2

14. Massie, M.L., Chun, B.N., Culler, D.E.: The ganglia distributed monitoring system: design, implementation, and experience. Parallel Comput. 30(7), 817–840 (2004)

15. Nagios (2010). http://www.nagios.org/

16. Newtouch (2015). http://www.newtouch.com/

17. Pareto (2000). http://en.wikipedia.org/wiki/pareto_principle

18. Rabkin, A., Katz, R.H. :Chukwa: a system for reliable large-scale log collection. In: USENIX Conference on Large Installation System Administration (LISA), vol. 10, pp. 1–15 (2010)

19. Rochwerger, B., Breitgand, D., Levy, E., Galis, A., Nagin, K., Llorente, I.M., Montero, R., Wolfsthal, Y., Elmroth, E., Caceres, J., et al.: The reservoir model and architecture for open federated cloud computing. IBM J. Res. Dev. 53(4), 535–545 (2009)

20. Zhu, Y., Hu, H., Ahn, G.-J., Yau, S.S.: Efficient audit service outsourcing for data integrity in clouds. J. Syst. Softw. 85(5), 1083–1095 (2012)

The Fault Tolerance of Big Data Systems

Xing Wu[1,2(✉)], Zhikang Du[1], Shuji Dai[1], and Yazhou Liu[2]

[1] School of Computer Engineering and Science, Shanghai University, Shanghai, China
{xingwu,duzhikang,daishuji}@shu.edu.cn
[2] Key Laboratory of Image and Video Understanding for Social Safety,
Nanjing University of Science and Technology, Nanjing, China
yazhouliu@njust.edu.cn

Abstract. When the size of the data itself becomes part of the problem, big data era is approaching. Big data technologies describe a new generation of technologies and architectures, designed to economically extract value from very large volumes of a wide variety of data, by enabling high-velocity capture, discovery, and/or analysis. Fault tolerance is of great importance for big data systems, which have potential software and hardware faults after their development. This paper introduces some popular applications and case studies of big data mining. The architecture of big data's individual components has parallel and distributed features, including distributed data processing, distributed storage and distributed memory, this paper briefly introduces Hadoop architecture of big data systems. Then presents some fault tolerance work recently in the big data systems such as batch computing, stream computing, Spark and Software defined networks, which shows great efforts to the capability of massive big data systems, and makes some comparison with each other.

Keywords: Fault tolerance · Big data · Hadoop · Spark · SDN

1 Introduction

Recently, big data becomes a highlighted buzzword in industry. Moreover, big data mining has almost immediately followed up as an emerging, interrelated research area. Databases, data warehouses, data marts and other information management technologies were about to solve the problem of large scale data.

However, "Big data" has become hot as an exclusive noun due to the rapid development of Internet, cloud computing, mobile and Internet of Things in recent years. Ubiquitous mobile devices, RFID, wireless sensors generate data all the time; hundreds of millions of users of Internet services always generate a huge amount of interactive data. The amount of data to be processed is too large, thus the business needs and competitive pressures require a real-time and effective data processing.

The traditional techniques cannot deal with such a large scale of data to satisfy the real business needs. Thus a number of new technologies have been developed and adopted, which includes distributed cache, distributed database, distributed file system, distributed storage scheme, no-SQL databases and so on. Among all these new technologies, fault tolerance is an inevitable part for big data systems.

J. Cao and J. Liu (Eds.): MiPAC 2016, CCIS 686, pp. 65–74, 2017.
DOI: 10.1007/978-981-10-3996-6_5

All big data systems need tolerate software and hardware faults remaining in the system after its development, which will benefit the systems in different ways including failure recovery, lower cost, improved performance and etc. A fault tolerance is a setup or configuration that prevents a computer or network device from failing in the event of an unexpected problem or error [1], as illustrated in Fig. 1. To make a computer or network fault tolerant requires users or companies to think how a computer or network device may fail and take steps that helps prevent that type of failure [2].

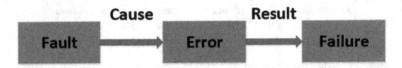

Fig. 1. Fault, error and failure

The paper is group by the following sections. Section 2 describes some popular applications of big data, including the applications in engineering, science, transportation and other fields. Section 3 discusses the technical architectures of big data systems, and uses Hadoop as an intuitive example. Section 4 summarizes the fault tolerant mechanisms and methods of some popular architectures of big data systems. Section 5 draws the conclusion of this paper.

2 Applications of Big Data

2.1 General Applications

Development of large data industry will promote the development of the world economy which has far-reaching significance from extensive to intensive changes to enhance the competitiveness of enterprises and the impact of the government's management capacity. Gathering the mass original data together, by the intelligent analysis and data mining technology, we analysis the potential of the data to forecast the development trend of the future things, helping people to make correct decisions to improve the operation efficiency of various areas and get greater benefits. Usually, the application of big data is the largest commercial areas, especially in electronic commerce. Its large market and its huge amount data from the market are particularly applicable to large data analysis techniques to predict and analysis to reduce costs, improve efficiency. In addition, the transportation, energy, electronics and other fields have conducted extensive research and application.

2.2 Case Studies of Big Data

There are some famous case studies listed in this paper.

Science Research. The data flow of the Large Hadron Collider (LHC) in experiments consists of 25 petabytes before replication and reaches up to 200 petabytes after replication.

Public Administration. The Obama administration project is a big initiative where a government is trying to find the uses of the big data which eases their tasks somehow and thus reducing the problems faced. It includes 84 different Big data programs which are a part of 6 different departments.

Business. Amazon uses data storage and processing 500,000 third-party electricity supplier sales data, optimizing the sales process and reduces costs.

Energy. With the analysis technology of big data, electrified wire net connects GPS and GIS and detects the fault of power transmission lines made by lightning and electrical traveling waves to make sure the reliability of electrical energy supply.

Traffic. Using of large data storage and analysis of the state of technology in aviation engine sensor data acquired thousands, such as Oil temperature, vibration frequency data, monitoring and forecasting the engine health condition, to ensure its reliable operation of the route running. Similar technology is also used in the automotive and motorcycle.

Health. The use of technology to monitor CT scanners and other large medical equipment status data, data processing a large number of sensors to ensure that services are available [2].

3   The Architecture of Big Data

Big data systems, have a basic part of the storage, processing, memory, network, etc. But based on the "4V" features of big Data technology, architecture of big data individual components has parallel and distributed features, including distributed data processing, distributed storage and distributed memory.

The big data platform Hadoop [3] utilizes MapReduce architecture to achieve a distributed data processing, in conjunction with HDFS distributed file system to achieve efficient, fault-tolerant and stable large data solution, as shown in Fig. 2. HDFS is a fault-tolerant and self-healing, distributed file system, the purpose of the standard server clustering into a large-scale expansion of the data pool. HDFS is the working load of large-scale data processing, engage in scalability, flexibility and throughput and specialized development. It accepts any data format for high bandwidth flow is optimized, can be extended to the environment in the 100 PB data over the deployment.

In HDFS, data replication in multi nodes will protect and maintain the computing performance. MapReduce is a highly scalable, parallel processing architecture, which is connected with HDFS and works together. The MapReduce and Hadoop, the calculation is executed in a data storage place, instead of moving the data to calculate the execution. The same physical data storage and computing nodes in the cluster on coexist. MapReduce can handle very large amounts of data, through the advantages of the nearest data, which is not the bottleneck bandwidth traditional limitation. The MapReduce will work load, divided into several parts and can be executed in parallel.

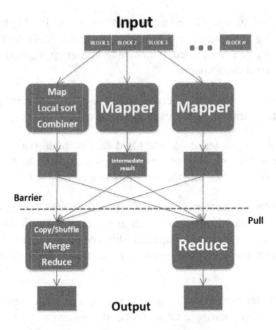

Fig. 2. The architecture of Hadoop

The traditional big data processing system use the disk to store the data. In recent years, with a substantial decline in the price of the hardware memory and real-time data processing based on WEB. In the face of big data processing needs more large-scale software vendors have introduced the database system based on their memory, such as HANA from and TimesTen from Oracle.

The same Spark big data platforms based on distributed memory also emerge as the times require. Big data platform Hadoop Spark and traditional distributed computing based on MapReduce is realized, but not for Hadoop Spark is to establish a set of distributed memory system, the task can be intermediate results stored in memory instead of distributed file system, and the need for iterative operation for distributed computing can achieve more efficient calculation. The core of Spark is to establish a flexible distributed data set RDD (Resilient Distributed Dataset). RDD is for distributed memory abstraction, RDD represents has been partitioned and can be operated in parallel data collection, RDD can be stored in memory, eliminating the need for large amounts of disk MapReduce operation. While its level of abstraction to avoid the direct operation of the memory, which can achieve the underlying hardware failures overshadowed by automatic reconfiguration for fault tolerance mechanisms.

4   Fault Tolerance in Big Data

The calculating pattern of big data is divided into batch computing and stream computing [4]. We first calculate the mass data storage, and then the static data of stored centralized computing. Hadoop is a typical batch of big data computing architecture, static data is responsible for the HDFS distributed file system storage, and is assigned by MapReduce computational logic to each data node for data calculation and value discovery. In stream computing, we cannot determine the arrival time and the arrival of the order of the data, all the data cannot be stored. Therefore, no streaming data storage, but when the flow of data in real-time arrival data is calculated directly in memory. Therefore, streaming data is no longer saved, but immediate real-time data is calculated in the memory when data of the flow arrival directly.

Such as Storm of Twitter [5], S4 of Yahoo, they are typical of the streaming data computing architecture, the data was calculated in the topology in task, and outputs the valuable information. Flow calculation and batch calculation are respectively applicable to different big data application scenarios: For some application scenarios, such as storage before computation, real-time demand not high, at the same time, the accuracy and completeness of the data being more important, batch computing mode is more suitable. For some application scenarios, such as without first storage, directly data calculation, real time being very strict, but the accuracy of the data requiring a little loose. Due to the application of the memory database, memory database fault tolerance has become a hot issue. Spark etc. are based on the calculation of distributed memory, using abstract RDD technique to realize fault tolerance mechanism to deal with memory fault of distributed memory in the system. In addition to the above two models of fault tolerance, a new method of fault tolerance appears recently, namely SDN (Software Defined Networking) fault-tolerant technology [6]. A level of abstraction will be established in the network hardware, such as routers, switches, gateway and other infrastructure, a virtual grid structure given software, namely SDN. All big data platforms run on SDN, while SDN is to deal with failure on the network. Thus the big data can be free from network fault tolerance, and we will focus on software failure.

4.1 Fault Tolerance in Batch Computing

In Hadoop batch calculation model, the applications of two main fault tolerant mechanisms to deal with failure are the data replication and the rollback mechanism [7].

The Data Replication. In data replication mechanism, a copy of the data will be in several different data nodes. When it needs data replication, any data node, that its communication is not the busy can copy data. The main advantage of this technology is that it can instantly recover from failure. But in order to achieve this kind of fault tolerance, storing data in different nodes will consume large amounts of resources, such as a waste of large amounts of memory and resources. When copying across different nodes, there is the possibility of data inconsistency. But the technology provides instantaneous fault recovery fast. Compared with the rollback method, this method is used more frequently.

Rollback Mechanism. The copy report will be saved in a fixed time interval. If failure occurs, the system is just to back up a save point, then starts the operation again from that point. The method adopts the rollback concept, that is, the system will return to the previous work. But this method increased the execution time of the whole system, because the rollback need to back up and to check on a save a consistent state, thus increasing the time. Compared with the first method, defects of the method is too time consuming, but needs less resources.

4.2 Fault Tolerance in Stream Computing

In stream computing system, there are four kinds of strategy to realize fault tolerance mechanism. Those are passive standby, active standby, upstream backup and a recent study of operator state management [8].

Passive Standby. System will be regular to back up the latest state on the master node to a copy of the replica node. When fault occurs, the system state will be restored from the backup data. Passive replication strategy support the case with data load higher, throughput larger, But the recovery time is longer, backup data can be saved in the distributed storage to reduce recovery time. This way is more suitable for precise data recovery, uncertainty computing applications. In the current, it is widely used in the calculation of streaming data.

Active Standby. When system transmits data for the master node, it also transmits a copy of the data for a copy of the node at the same time. When the master node failed, a copy of the node completely takes over the work, and the deputy nodes need to assign the same system resources. Fault recovery time is shortest in this way, but smaller data throughput.it also wastes more system resources.

Upstream Backup. Every master node is recording its own state and the output data to a log file. When a master node failed, the master node of upstream will replay a copy of the data in a log file to the corresponding node, for recalculating the data. They need longer time to reconstruct the state of the recovery, so that fault recovery time tends to be long. As system resources are scarce, upstream backup strategy is a better option in a state of the circumstances of fewer operators.

Operator State Management. This method will display the state operator to stream processing system through state management main types, And according to the time interval, it will back up its state to the upstream nodes. In any malfunction, the system will spread laterally to replace the fault node, and recover from the upstream node state. This method does not need longer reconstruction time, but also take up less computing resource.

4.3 Fault Tolerance in Spark

Compared with the parallel file system, the big data system using memory access speed, based on distributed memory, has more efficient data processing ability. Its face fault-tolerant mechanism will be different due to different ideas of this. For example, the RDD memory abstraction technique was used to realize fault tolerance mechanism for Spark framework [9].

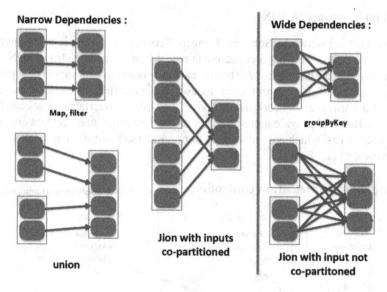

Fig. 3. Two types of dependencies in Spark

RDD (Resilient Distributed Dataset) stored all calculation data in a Distributed memory. RDD stored data in the cluster of memory by partition way. There are two kinds of operation: The Transformation and The Action. Transform is the operation that a kind of RDD is converted to another RDD, similar to Hadoop Map operation, whose operator definition is rich, including map, join, filter, groupByKey operations and so on. Action is similar to Reduce in Hadoop. The output is an aggregation function values such as the count, or a collection [10].

The Spark uses two methods to realize fault tolerance. One method is traditional checkpoint, which is to restore the backup RDD data set. This method will store RDD in file system on a regular basis, similar to Hadoop. Another method is to use lineage to realize fault tolerance mechanism. This method is used to establish the Transformation operation of a data set, used for data recovery. It is similar to the upstream backup strategy in the loss calculation system, its application range depending on the dependent type. As shown in Fig. 3, the Spark of the two types of dependence, Narrow dependence and upstream and downstream RDD is saved in the same node. So the node downtime cannot continue to back up. In wide dependence, upstream and downstream RDD is saved on different nodes. When the downstream is down, it can be recovery through the upstream. Different operations (union, map, join, etc.) will produce different depend on the type. There are two kinds of fault tolerance mechanism in the Spark. It depends on the case. The both kinds of recovery methods used in distributed memory system need to consume a lot of time Due to the attention of the computational efficiency, there is the corresponding loss on the efficiency of the fault recovery.

4.4 Fault Tolerance in SDN

Software Defined network (Software Defined Networking, SDN) is a revolutionary change in the field of network architecture in recent years. The core idea of SDN is that the data layer and the control layer of traditional network equipment should be separated, and the function of the control layer focuses on controllers. We can manage and configure a variety of network devices via Standardized interfaces in a centralized controller. That will provide more possibilities for the design, management and use of network resources, which are more likely to promote the innovation and development of the network [11] (Fig. 4).

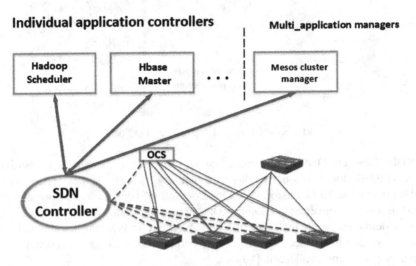

Fig. 4. The big data applications based on SDN

From a virtual data center level, SDN centralized master data center network topology information of control surface, and that the virtual machine's MAC/IP/location attribution can realize flexible programmable ability, and can keep fault in physical layer away from the application layer. That will provide a more flexible and more stable and reliable virtual network for network virtualization, automation, all sorts of network services and big data.

Big Data applications are no longer connected through the traditional network hardware and provide services, but by SDN control operation in the virtual network. When network hardware has the failure, big data applications should keep away from hardware fault by SDN controller. SDN is responsible to maintain the availability of virtual network by calling the backup hardware. So that big data applications don't have to focus on network level fault, and only focus on the software failure.

5 Conclusion

Throughout the past 20 years, development of IT technology and IT industry, based not only on the progress and development of the technology itself, contains more business model and application of the success of the marketing mode. Big data has been deeply rooted in every area of life and technology, and it has made a great contribution for business, engineering, health care, public administration. With commonly used Hadoop platform as an example, this paper discussed the basic architecture of big data technology, and provides big data applications in various industries. While fault tolerance is an important issue in big data applications, fault-tolerant mechanism ensures that large data software and applications can run stably and reliably, and continue to provide service for various industries. This paper summarizes the technology of data in a variety of fault tolerant mechanism, and provides the comparison. It also provides reference of high availability and high reliability for big data applications.

Acknowledgements. This paper is supported by the project 61303094 supported by National Natural Science Foundation of China, by the Science and Technology Commission of Shanghai Municipality (16511102400), by Innovation Program of Shanghai Municipal Education Commission (14YZ024).

References

1. Jhawar, R., Piuri, V., Santambrogio, M.: A comprehensive conceptual system-level approach to fault tolerance in cloud computing. In: 2012 IEEE International Systems Conference (SysCon), pp. 1–5. IEEE (2012)
2. Dyavanur, M., Kori, K.: Fault tolerance techniques in big data tools: a survey. Int. J. Innovative Res. Comput. Commun. Eng. 2(2), 95–101 (2014)
3. Parker, P.A.: Discussion of "reliability meets big data: opportunities and challenges". Qual. Eng. 26(1), 117–120 (2014)
4. Shvachko, K., Kuang, H., Radia, S., et al.: The hadoop distributed file system. In: 2010 IEEE 26th Symposium on Mass Storage Systems and Technologies (MSST), pp. 1–10. IEEE (2010)
5. Neumeyer, L., Robbins, B., Nair, A., et al.: S4: distributed stream computing platform. In: 2010 IEEE International Conference on Data Mining Workshops (ICDMW), pp. 170–177. IEEE (2010)
6. Jones, M.T.: Process real-time big data with Twitter Storm. IBM Tech. Libr. 14(2), 1–5 (2013)
7. Reitblatt, M., Canini, M., Guha, A., et al.: Fattire: declarative fault tolerance for software-defined networks. In: Proceedings of the Second ACM SIGCOMM Workshop on Hot Topics in Software Defined Networking, pp. 109–114. ACM (2013)
8. Antoniu, G., Costan, A., Bigot, J., et al.: Scalable data management for map-reduce-based data-intensive applications: a view for cloud and hybrid infrastructures. Int. J. Cloud Comput. 2(2), 150–170 (2013)
9. Hwang, J.H., Balazinska, M., Rasin, A., et al.: High-availability algorithms for distributed stream processing. In: Proceedings of 21st International Conference on Data Engineering 2005, ICDE 2005, pp. 779–790. IEEE (2005)
10. Zaharia, M., Chowdhury, M., Das, T., et al.: Resilient distributed datasets: a fault-tolerant abstraction for in-memory cluster computing. In: Proceedings of the 9th USENIX Conference on Networked Systems Design and Implementation, p. 2. USENIX Association (2012)

11. Zaharia, M., Chowdhury, M., Franklin, M.J., et al.: Spark: cluster computing with working sets. In: Proceedings of the 2nd USENIX Conference on Hot Topics in Cloud Computing, p. 10 (2010)
12. Kim, H., Santos, J.R., Turner, Y., et al.: Coronet: fault tolerance for software defined networks. In: 2012 20th IEEE International Conference on Network Protocols (ICNP), pp. 1–2. IEEE (2012)

A Market-Based Analysis of Bidding Strategy Between Web Service Providers and Users

Bing Shi[✉], Zhaowei Wang, and Guangyi Hu

School of Computer Science and Technology,
Wuhan University of Technology, Wuhan, China
bingshi@whut.edu.cn

Abstract. In the cloud environment, there exist multiple providers offering the same or similar web service, and multiple users requiring the same web service. There exist competition among web service providers and users. In this paper, we investigate the interacting strategy between web service providers and users based on the double auction mechanism. In this setting, web service is traded as commodity between service providers (sellers) and users (buyers). Web service providers and users interact with each other, and they need to submit effective offers for the traded web service. We then use game theory to analyze how web service providers and users bid in different trading environments with different budget constraints. We find that if one-unit service is allowed in the market-place, service users shade (i.e. bid less than their types) their bids less and service providers shade (i.e. ask more than their types) their asks more when the budget increases. If multi-unit services are allowed in the marketplace, when the service providers' budgets increase, service providers shade their asks less and service users shade more. In addition, more service users are willing choose to enter the market. When the service users' budget increases, more and more users offer two-units services to obtain more profits. Our results will provide guidance for the efficient design of bidding algorithms between web service providers and users.

Keywords: Web service providers and users · Double auction mechanism · Game theory · Bidding strategies

1 Introduction

Now various types of web service exist in the Internet. As the development of web technology, a number of web service providers provide the same or similar web service while a number of service users seek the same or similar web service. Therefore an efficient interaction mechanism is required to deal with the interaction between service providers and users. The web service providers and users in the Internet usually belong to different self-interested organization, and they intend to maximize their own profits. Web service users can use the service only after paying to providers. Furthermore, the service quantity which the service providers can provide is limited while the budget that the service users can use to pay is also limited. Web service providers and users should consider this constraint during the interaction. Therefore, we will analyze how

© Springer Nature Singapore Pte Ltd. 2017
J. Cao and J. Liu (Eds.): MiPAC 2016, CCIS 686, pp. 75–86, 2017.
DOI: 10.1007/978-981-10-3996-6_6

the service providers and users interact with each other to allocate web services between them under the budget-constraint.

Actually, we can see that the exchange of web service between service providers and users is similar to the commodity exchange between buyers and sellers in the market. Therefore, in this paper, we assume that web service users are buyers, web service providers are sellers, and web services are commodities. The interaction between web service providers and users can be considered as the matching between sellers and buyers in the market. Since in this setting, we have multiple service providers and users for the same web service, the double auction mechanism, which allows multiple buyers and sellers to trade simultaneously, is a suitable model [1]. In such a mechanism, the traders can submit offers at any time, and the market matches the buyers and sellers at a specific time. The market determines the actual transaction price for the matched buyers and sellers based on the pricing policy. In this situation, traders need effective bidding strategies to decide how much to offer for the commodity to maximize their profits.

There exist a lot of works on the interaction about web services. Perry notes the impact of cognitive level on Web service interaction [2]. Lehman considers web services interaction as phenomena and actions, extracts several laws and gives the relevant model [3]. Simon has put forward a new point of view and believes that the web services interaction can be seen as a combination point of internal and external environment [4]. However, these works have not considered that service providers and users are selfish and they intend to maximize their profits during the interaction. Since this paper will analyze how service users and providers interact by bidding effectively, we also need to discuss the existing trading strategies in the double auction market. In 1994, Rustichini studied the equilibrium quotes of traders in the double auction market and their deviations from the actual valuation of the goods [5]. They proved that the largest deviations in a large market were also small. This conclusion fully demonstrates the validity of this mechanism in the double auction market and the feasibility of using it as a theoretical model to analyze the problem. The researchers also studied how to bid in the double auction market and designed various algorithms such as GD [6], ZIP [7]. However, these works are heuristic-based and there is no theoretical answer how traders bid in the market. In addition, these algorithms do not take into account the budget constraints of the traders.

In this paper, we will analyze the bidding strategy of web service providers and users with budget-constraint based on the double auction mechanism. We regard service providers as sellers, service users as buyers, web services as commodities. Intuitively, the bidding strategies of web service providers and users are affected by each other. Therefore, we will use game theory to analyze their strategies. Specifically, we use fictitious play algorithm to compute the Nash equilibrium bidding strategy of service providers and users under different budget constraints. Our results will provide guidance for the efficient design of bidding algorithms between web service providers and users.

The structure of the rest of the paper is as follows. In Sect. 2, we introduce the market model which service providers and users interact with each other. In Sect. 3, we describe the fictitious play algorithm which will be used to compute the Nash equilibrium. In Sect. 4, we analyze the equilibrium strategy in different trading environments under different budget constraints. Finally, we conclude in Sect. 5.

2 Market Model

In this section, we introduce the market model for the web service providers and users.

2.1 Basic Setting

We assume that there is a set of users $B = \{1, 2, 3 \ldots B\}$, and a set of providers $S = \{1, 2, 3 \ldots S\}$. Each web service provider or user has a reserved price which is defined as a type $\theta, \theta \in [0, 1]$. For service users, the type can be considered as the highest price that a service user is willing to pay for a web service; for service providers, the type can be regarded as the lowest price that a service provider agrees to sells a web service [8]. In the Internet environment, the types of web service providers and users usually are private knowledge, which means others do not know this information. However, the type probability distribution function is public. We assume that the distribution functions of the users and the providers are F^B and F^S respectively, and the corresponding probability density functions are f^b and f^s respectively.

In the web service market, a service user's offer is called a bid, a service provider's offer is called an ask. The bidding action is a tuple $\sigma = (n, d)$ where n is the quantity of web services traded and d is the offer that the service providers or users submit. This tuple action means that the service provider or user is willing to sell or buy n web service instances at price d. n is drawn from a discrete set $N = \{1, 2 \ldots N\}$. d is drawn from a discrete set $D = \left\{ 0, \frac{1}{D}, \frac{2}{D} \ldots \frac{D-1}{D}, 1 \right\} \cup \{\emptyset\}$ (means not choosing the market-place). The action space is $\Gamma = N \times D$. In addition, the service users have the limitation on the budget, and service providers have the limitation on the number of service instances. We assumed that service users' budget is BL and service providers' budget is SL, so service users' action space is $\Gamma^b = \{\sigma^b = (n, d) \in \Gamma : n \times d \leq BL\}$ and service provider's action space is $\Gamma^s = \{\sigma^s = (n, d) \in \Gamma : n \leq SL\}$. We use $\Omega^b = \left\{ \omega_1^b, \omega_2^b, \omega_3^b \ldots \omega_{|\Gamma^b|}^b \right\}$ to denote the probability of each action being chosen by the service users under the budget BL. ω_i^b is the probability that service user adopts action. Similarly, for the service providers, we use probability distribution Ω^s to represent the probability of each action selected.

In this paper, we assume that the service market adopts the equilibrium matching rule and the k pricing-policy. Equilibrium matching rule is to match the highest bid of the service user with the lowest ask of the service provider [9]. The pricing policy is that the transaction price of the service user and the service provider is determined by the pricing parameters k, within the matched bid and ask [10].

2.2 The Expected Utility

In this section, we introduce how to compute a service user's expected utility when adopting a specific action. The expected utility of service provider can be calculated similarly. We set a service user's action is σ and the type is θ, then its expected utility is denoted by $U(\Omega^b, \Omega^s, \theta, \sigma)$. Suppose that x_i is the number of remaining users using action σ_i^b and y_i is the number of providers using action σ_i^s. We use $x = \left\langle x_1, x_2, x_3 \ldots x_{|\Gamma^b|} \right\rangle$, $\sum_{i=1}^{|\Gamma^b|} x_i = B - 1$ and $y = \left\langle y_1, y_2, y_3 \ldots y_{|\Gamma^s|} \right\rangle$, $\sum_{i=1}^{|\Gamma^s|} y_i = S$ to represent the number of each action being chosen by users and providers respectively. Tuples x, y belongs to sets X, Y respectively. X, Y are the sets of all such possible tuples. The expected utility function can be expressed as:

$$U(\Omega^b, \Omega^s, \theta, \sigma) = \sum_{x \in X, y \in Y} p(x) \times p(y) \times U(\theta, \sigma, x, y) \tag{1}$$

where $p(x)$ and $p(y)$ represent the probability of occurrence of tuple x and y respectively and $U(\theta, \sigma, x, y)$ represents the expected utility of the service user under tuple x and y. Furthermore, we have:

$$p(x) = \binom{B-1}{x_1, x_2, x_3, \ldots x_{|\Gamma^b|}} \times \prod_{i=1}^{|\Gamma^b|} \left(\omega_i^b \right)^{x_i} \tag{2}$$

$$p(y) = \binom{S}{y_1, y_2, y_3 \ldots y_{|\Gamma^s|}} \times \prod_{i=1}^{|\Gamma^s|} \left(\omega_i^s \right)^{y_i} \tag{3}$$

The expected utility of the service user is also related to the position of the action's bid in the service market. Therefore, we also discuss the ranking of the action's bid in the service market. For convenience, we first define the following two functions.

$$h(x, d(\sigma)) = \sum_{\sigma_i^b \in \Gamma^b : d(\sigma_i^b) > d(\sigma)} x_i \times n(\sigma_i) \tag{4}$$

where $h(x, d(\sigma))$ indicates the number of action's bids that are higher than the bid d of action s when other users' actions are satisfied for tuple x.

$$e(x, d(\sigma)) = \sum_{\sigma_i^b \in \Gamma^b : d(\sigma_i^b) = d(\sigma)} x_i \times n(\sigma_i^b) \tag{5}$$

where $e(x, d(\sigma))$ indicates the number of action's bids that are same as the bid d of action σ when other users' actions are satisfied for tuple x.

Assuming that x bids are the same as the bid d of action σ, and the probability that each bid appears in the x-th position is equal when the service market matches the x bids.

$$U(\theta,\sigma,x,y) = \frac{n(\sigma)}{e(x,d(\sigma)) + n(\sigma)} \times \sum_{v=h(x,d(\sigma))+1}^{h(x,d(\sigma)) + e(x,d(\sigma)) + n(\sigma)} U(\theta, d(\sigma), v, y) \quad (6)$$

where $U(\theta, d(\alpha), v, y)$ represents the expected utility when the action' bid of the service user are ranked v in the service market. Let $vth(y,v)$ be the vth ask when the service provider's action is satisfied with tuple y. Eventually, we have:

$$U(\theta, d(\sigma), v, y) = \begin{cases} \theta - vth(y,v) \times k - d \times (1-k) - \lambda, & d(\sigma) \geq vth(y,v) \\ -\lambda(\text{no transaction}), & d(\sigma) < vth(y,v) \end{cases} \quad (7)$$

where λ is a small fee to be paid when service users and providers enter the service marketplace.

3   The Fictitious Play Algorithm

Fictitious Play algorithm is a computational learning algorithm, and can be used to calculate the Nash equilibrium. In the standard FP algorithm, the participants are assumed to use a mixed strategy. The participants observe the using frequency of each strategy in the offer space, and estimate the mixed strategy. In the FP algorithm, the observed frequency of each strategy is called the FP belief. In each round, each participant evaluates the mixed strategy of the remaining participants by observing the history of the game, and then calculates the best response strategy, and finally updates the FP beliefs. All participants iterate the process until the algorithm converges. However the standard FP algorithm is not suitable for our setting where the type of web services users and providers is private knowledge since we do not know which type perform which action, and thus cannot estimate the mixed strategy. In [11], a generalized fictitious play algorithm was proposed to analyze the games with continuous types and incomplete information. Based on this algorithm, if the players' action space is finite, the FP beliefs will converge to a ε-Bayes−Nash equilibrium.

3.1   Computing the Best Response

In Sect. 2.1, we use Ω^b and Ω^s to denote the probability distributions of users' and providers' action respectively. In this section, we use them to represent FP beliefs about users' and providers' action respectively. Given the FP beliefs, the best response action should maximize the utility of the web service users and providers. We compute the users' best response function as follows:

$$\sigma^{b*}(\Omega^b_\tau, \Omega^s_\tau, \theta^b) = \arg\max_{\sigma \in \Gamma^b} U^b(\Omega^b_\tau, \Omega^s_\tau, \theta^b, \sigma) \quad (8)$$

Where $\sigma^{b*}(\Omega^b_\tau, \Omega^s_\tau, \theta^b)$ represents the best response function for service users of type θ^b with probability distributions $\Omega^b_\tau, \Omega^s_\tau$ for other service users and services providers.

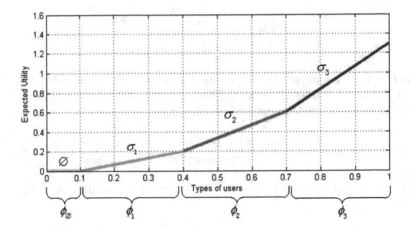

Fig. 1. Piecewise linear optimal utility function

With the best response function, we can calculate the optimal expected utility function. We note that the users' expected utility function is linear in its type for a given action. In this paper, since the number of action is finite, the optimal expected utility function is the upper envelop of a finite set of linear function, and thus is piecewise linear. An example with 4 action $(\sigma_1, \sigma_2, \sigma_3, \emptyset)$ is shown in Fig. 1. The optimal utility function is a piecewise linear function and each line segment $\sigma_i(\emptyset)$ corresponds to a type interval $\phi_{i(\emptyset)}$ on the x-axis.

3.2   Updating the FP Beliefs

With the type intervals corresponding to the best response, we can compute the probability distributions of each action under the current FP beliefs, and then update the FP beliefs of the next round. We denote ϖ_i^b as the probability that service users choose the best response action σ_i^b corresponding to type interval ϕ_i^b, so $\varpi_i^b = \int_{\phi_i^b} f^b(\theta)d\theta$. In the τth round, the best response distribution of service users is represented by $\Omega^b = \left\{\varpi_1^b, \varpi_2^b, \varpi_3^b \dots \varpi_n^b\right\}$. Given the FP beliefs of current iteration round Ω_τ^b, we updated the FP belief: $\Omega_{\tau+1}^b = \frac{\tau}{\tau+1} \times \Omega_\tau^b + \frac{1}{\tau+1} \times \Omega^b$, where $\Omega_{\tau+1}^b$ is the updated FP belief for the next iteration round $\tau+1$. The computation of the providers' best response function and the belief updates is analogous.

3.3   Measuring Convergence

In our model, we intend to analyze service providers and users' $\varepsilon-$Nash equilibrium bidding strategies. This means that, if the absolute value of difference of estimated utility in two rounds is less than a small value ε, the algorithm convergence. The measure of convergence is as follows:

$$\left|\tilde{U}^b(\Omega^b_\tau, \Omega^s_\tau) - \tilde{U}^b(\Omega^b_{\tau-1}, \Omega^s_{\tau-1})\right| \leq \varepsilon \text{ and } \left|\tilde{U}^s(\Omega^b_\tau, \Omega^s_{\tau-1}) - \tilde{U}^s(\Omega^b_\tau, \Omega^s_{\tau-1})\right| \leq \varepsilon \quad (9)$$

where $\tilde{U}^b(\Omega^b_\tau, \Omega^s_\tau)$ is the expected utility of users adopting the best response action against the current iteration rounds' FP beliefs:

$$\tilde{U}^b(\Omega^b_\tau, \Omega^s_\tau) = \int_0^1 U(\Omega^b_\tau, \Omega^s_\tau, \theta, \sigma^b) \times f^b(\theta) d\theta \quad (10)$$

where σ^b is the best response action of the users with type θ given FP beliefs Ω^b_τ and Ω^s_τ. $\tilde{U}^b(\Omega^b_{\tau-1}, \Omega^s_{\tau-1})$ is the expected utility of users adopting the best response action against the last iteration rounds' FP beliefs. The equations for providers are analogous.

3.4   Algorithm Overview

Based on the above description, an overview of FP algorithm is shown in Fig. 2.

Initial:

Set the initial FP belief Ω^b_τ, Ω^s_τ, set the iteration count $\tau=0$.

do

1. Calculate best response function: $\sigma^{b*}\left(\Omega^b_\tau, \Omega^s_\tau, \theta^b\right)$ and $\sigma^{s*}\left(\Omega^b_\tau, \Omega^s_\tau, \theta^s\right)$.

    Generate the type interval ϕ^b_i corresponding to the best response action σ^b.

    Generate the type interval ϕ^s_i corresponding to the best response action σ^s

2. Compute inherent action distribution Ω^b, Ω^s of users and providers.

3. Update FP belief $\Omega^b_{\tau+1}$ and $\Omega^s_{\tau+1}$,

4. Measure the convergence, If convergence, return $\Omega^b_{\tau+1}, \Omega^s_{\tau+1}$

5. Set $\tau = \tau+1$

while (!converged)

Fig. 2. The fictitious play algorithm

4   Equilibrium Analysis

In this section, we analyze the Nash equilibrium bidding strategies of web service users and providers. In Nash equilibrium, no one can gain more profits by individually deviating its current strategy. In the following analysis, for illustrative purpose, we consider 10 service providers and 10 service users, and set $D = \{0, 0.1, 0.2, 0.3, 0.4, 0.5, 0.6, 0.7, 0.8, 0.9, 1.0\} \cup \{\emptyset\}$. We consider the uniform distribution for types and initial FP beliefs of web service providers and users. We set the pricing parameters $k = 0.5$. In addition, we set $\varepsilon = 0.00001$ in the ε-Bayes-Nash equilibrium. We also assume that the small cost for users and providers to enter the marketplace is $\lambda = 0.0001$. In the following analysis, we will consider different budget constraints which allow web service users and providers to trade one-unit services or multi-unit services.

4.1   For One-Unit Service Marketplaces

In the one-unit service marketplace, each service provider and user can only trade one-unit web service instance. We first analyze this simple case, which can guide the following analysis of trading multi-unit service. Figure 3 shows the equilibrium strategies for service users and providers when service users' budget is 0.4, 0.6, 0.7 and 1 respectively. When users' budget is less than or equal to 0.7, we find that service users intend to shade (i.e. bid less than their types) their bids less and service providers intend to shade (i.e. ask more than their types) their asks more when the service users' budget increases. As shown in Fig. 3(a), when the users' budget is 0.4, service users with the type 0.6 bid 0.4 and service providers with the type 0.4 ask 0.4, but when the users' budget is increased to 0.6 (Fig. 3(b)) service users and service providers' offer are changed to 0.5. Since when users' budget increases, the service users will raise their bids to raise the matching probability in the service market in order to improve their expected income, and service providers raise their asks can improve their own transaction price, and thus improve their expected earnings. In addition, the largest type of service providers entering the market is always equal to the budget. This is because when service providers' type is larger than the users' budget, service providers cannot gain positive revenue, and will not enter the marketplace. As the budget increases, the range of types of service providers entering the market increases. As shown in Fig. 3 (a), when the users' budget is 0.4, service providers within the range of type [0, 0.4] chooses to enter the market, and as shown in Fig. 3(b)), when users' budget is 0.6, the type's range of service providers entering the market increases to [0, 0.6]. This is because when users' budget increases, high-type service users improve their bids in order to reach a deal and thus high-type service providers can get the benefits and intend to enter the market. After the budget is greater than or equal to 0.7, the action choice of the service users and providers is not changed, because the service users will also shade bids in order to guarantee their expected income. Since service users and service provider's offers are affected by each other, the service user's bid does not change, and the corresponding service provider's asks will no longer change as well.

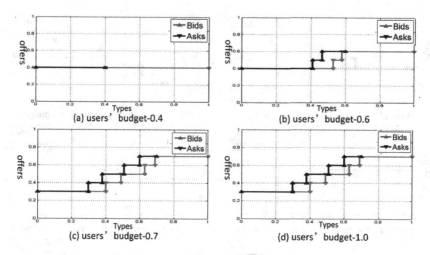

Fig. 3. Equilibrium strategies in one-unit service marketplaces

4.2 For Multi-unit Service Marketplaces

We now extend the above analyze to the case with multi-unit services where web service providers and users are allowed to trade multi-unit service instance. This is more closely to the realistic setting. Here we set the maximum number N of allowed as 2. In the multi-unit service marketplaces, there are two types of budgets constraint: limited budgets for the service users and limited allowable number of service instance for the service providers. Figure 4 shows equilibrium bidding strategy when service user's budget is limited by 0.5, 1.0, 1.5 and 2.0 respectively and service providers' allowable number is limited 1 and 2 respectively. Note that when service users or service providers' budget is 0, users and providers do not enter the market because of the small cost λ.

We first analyze the change of the trading strategy when service users' budget is same and service providers' budget is different. When the service provider's budget is from 1 to 2, the service providers intend to shade their asks less, service users intend to shade more. As shown in Fig. 4(a), service users with type 0.4 and service providers with type 0.4 bid 0.4, whereas service providers and the users' offer has changed to 0.3 in Fig. 4(b). In addition, the range of type that the service users choose to enter the market increases. In Fig. 4(c), the range of type of service users entering the market is [0.5–1], and in Fig. 4(d) this is changed to [0.4–1]. This is because the number of services instance that the service provider can provide increases as the service providers' budget is 2. This results in severe competition among service providers. Service provider will reduce their asks in order to increase the probability of successful matching. The corresponding competition between services users reduces, and they intend to lower their bids to increase the expected income. In this case, low-type service providers can also get benefits, and they are willing to enter the market.

Next, we analyze the change of the trading strategy when service providers' budget is same and service users' budget is different. We find that, as well as one-unit service

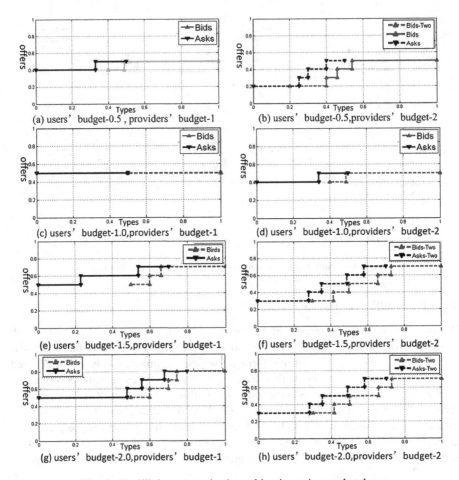

Fig. 4. Equilibrium strategies in multi-unit service marketplaces

marketplace, the servers' strategy does not change when the users' budget is greater than a value in the multi-unit service marketplace. The value is 1.6 when the service providers' budget is 1, and decreases to 1.4 when the service providers' budget is 2. This is because the value is also related to the competition between users and providers. When the service providers' budget is 2, there would be more available service in the market and service users could successful match service with higher probability. In this situation, service users intend to shade their bids more, and thus the budget' influence is more quickly lost and the value is decreased. When users' budget is less than the value, the changes of users and providers' offers can be divided into three stages. We introduce the case that providers' budget is 1, and the case that providers' budget is 2 is similar. The first stage is [0, 0.7], the service users' actions only have one-unit service, and they shade their bids less with budget increase. [0.7, 1] is the second stage. When users' budget increases, high-type service users shade their bids more, low-type users shade their bids less, and more and more users offer two-units services. And the type's range of users entering the marketplace also becomes smaller. Providers shade their asks less and their maximum ask becomes smaller. [1.0, 1.6] is the third stage, all service users offer two-units services, but users shade their bids less with the users' budget increases. In addition when the service provider's budget is changed to 2, all three stages are advanced [0, 0.5], [0.5, 0.8], and [0.8, 1.4]. This is because when service providers' budget increases, the competition among service providers becomes larger, and thus the competition among service users becomes smaller, which led to the advance of the three stages.

5   Conclusion

In this paper, we analyze how web service providers and users bid for the web service in different marketplace situations. We use a market-based mechanism to model this problem, and use the Fictitious Play algorithm to analyze the equilibrium bidding strategy of web service providers and users. We obtained a number of useful insights. We found that if one-unit service is allowed in the marketplace, service users shade their bids less and service providers shade their asks more when the service users' budget increases. If multi-unit services are allowed in the marketplace, when the service provider's budget increases, the service providers shade their asks less, and service users shade more. In addition, the range of type that the service users choose to enter the market increases. When the service users' budget increases, more and more users offer two-units services to obtain more profits.

Acknowledgments. This paper was funded by the National Natural Science Foundation of China (No. 61402344), Scientific Research Foundation for the Returned Overseas Chinese Scholars, Ministry of Education of China.

References

1. Friedman, D., Rust, J.: The Double Auction Market: Institutions, Theories and Evidence. Santa Fe Institute Studies in the Science of Complexity, vol. XIV. Perseus Publishing, Cambridge (1993)
2. Perry, D.E.: Dimensions of software evolution. In: Proceedings of International Conference on Software Maintenance 1994, pp. 296–303. IEEE, September 1994
3. Lehman, M.M., Ramil, J.F., Kahen, G.: Evolution as a noun and evolution as a verb. In: SOCE 2000 Workshop on Software and Organisation Co-evolution, vol. 9, p. 31, July 2000
4. Simon, H.A.: The Sciences of the Artificial. MIT Press, Cambridge (1969)
5. Rustichini, A., Satterthwaite, M.A., Williams, S.R.: Convergence to efficiency in a simple market with incomplete information. Econometrica: J. Econometric Soc. **62**, 1041–1063 (1994)
6. Cliff, D., Bruten, J.: Minimal-intelligence agents for bargaining behaviors in market-based environments. Hewlett-Packard Labs Technical Reports (1997)
7. Gode, D.K., Sunder, S.: Allocative efficiency of markets with zero-intelligence traders: market as a partial substitute for individual rationality. J. Polit. Econ. **101**, 119–137 (1993)
8. Mas-Collel, A.: Microeconomic Theory. Oxford University Press, Oxford (1995)

A Lightweight Hash-Based Mutual Authentication Protocol for RFID

Zhangbing Li[1,2(✉)], Xiaoyong Zhong[1],
Xiaochun Chen[1], and Jianxun Liu[1,2]

[1] School of Computer Science and Engineering,
Hunan University of Science and Technology, Xiangtan, Hunan, China
lzb_xt@126.com, chen_xiaochuns@126.com,
zxyhnust@163.com, jx529@gmail.com
[2] Key Lab of Knowledge Processing and Networked Manufacturing,
College of Hunan Province, Xiangtan, Hunan, China

Abstract. For the RFID authentication protocols based on Hash functions, there are some shortcomings, such as imperfect defense on the attacks, intensive calculation, time-consuming authentication process, and so on. By using of the dynamic-shared key and one-way feature of Hash function, a lightweight Hash-based mutual authentication protocol has been proposed and proved by SVO logic in this paper. It avoids an exhaustive search in the back-end database, and supports the transfer of ownership of the tag and the scalability of RFID system. Besides resisting the common attacks, the protocol is suitable for the RFID system that needs to be low-cost, lightweight computing and large numbers of tags, which is of significant merit for RFID application.

Keywords: RFID · Mutual authentication · Hash function · SVO logic · Security protocol

1 Introduction

In recent years, radio frequency identification (RFID) technology has been widely used in supply chain management, target detection and tracking, electronic payment, environmental monitoring and so on, but the information leakage and other security issues are also increasingly highlighted [1]. A complete RFID system is composed of three parts: a reader, tags and back-end database. It may suffer from the main attacks as follows: message replay, denial of service, tag clone camouflage, reader fake, unauthorized access and track, desynchronization etc. Therefore a RFID system needs a strong security protocol that can withstand the above attacks to meet the security demands and protect the data privacy. RFID authentication is designed to make mutual authentication between the reader and tag, but does not allow any attacker to recover the intimate information in the course of authentication process. Thus, to design a secure authentication protocol is becoming a research hotspot [2].

Authentication protocols need to encrypt the sensitive data for preventing information leakage or tampering with hackers. Encryption function should be able to ensure the data integrity and guarantee the confidentiality and factuality of RFID

© Springer Nature Singapore Pte Ltd. 2017
J. Cao and J. Liu (Eds.): MiPAC 2016, CCIS 686, pp. 87–98, 2017.
DOI: 10.1007/978-981-10-3996-6_7

system. In view of the limited resources and storage space in RFID tag, the RFID authentication process will become very difficult since some encryption algorithm has a complicated computation in the practical application. Hash function is a frequently used algorithm for the RFID protocol with reliable safety and acceptable computing cost [1, 2]. Some typical authentication protocols based on hash function are: random Hash-Lock protocol, Hash-Chain protocol, RFID Library protocol, LCAP (Low-cost RFID Authentication Protocol) [3–9], and some improved protocols based on them [1, 2, 10–20]. These protocols all assume that the channel between the back-end database and the reader is secure but insecure between the reader and Tags, and requires authentication.

In 2002 Sarma [3] proposed Hash-Lock protocol which uses metaID(equal to hash (key)) to replace the true ID of the tag to protect the data privacy. But it will easily suffer from replaying and spoofing attack since the metaID value is unchanged during each communication, and the protocol also does not prevent tracking. Then Weis [4] mended the Hash-Lock protocol using the unpredictability of the random number. This protocol is called RHL protocol which ensures the indistinguishability of the session data and resists the position tracking, but the plaintext transmission for the tag ID still does not resist the counterfeiting and replay attacks. The Hash-Chain protocol proposed by Ohkubo [5] uses two different Hash functions to refresh the ID dynamically, and is also with strong ability of anti tracking. But it can only achieve a one-way authentication, and is vulnerable to the man-in-the-middle and replay attacks. Henrici [6] proposed a protocol based on hash ID-changed, which introduces the identification information to prevent man-in-the-middle attack, but there are some risks of desynchronization between the tag and database. The LCAP protocol based on distributed inquiry-response mode is proposed by Rhee et al. [7], which imports the random number in both the reader and tag. But there exists of the problem for forward security if the attacker gets the tag ID, and also the risk of losing synchronization between the tag and database. Molnar et al. [8] proposed David digital library protocol, which is a mutual authentication protocol and different from the hash ID-changed protocol. It makes use of a static ID and the shared secret value S to achieve the authentication between the server and tag, but the authentication is time-consuming and has of intensive calculation and high cost. In 2006, Tsudik [9] proposed the YA-TRAP's authentication protocol which introduces the time stamp, but it is vulnerable to denial of service attacks (unable to distinguish from illegal and legal tags).

Some domestic scholars have also made the design and improvement of RFID security protocols [10–17]. Li [10] introduces the random number to prevent replay attacks in the improved Hash-Chain protocol, but it will appear Dos attack when the number of illegal tags is more than of $(M + T - 1)$. Hash-Chain protocol based on two-dimensional interval are proposed by Xiong [11], which increases the index (Ai, Bi) for each tag, but there are threats of replay attack and impersonation attack. Yuan improved protocol [12] hides the tag's ID for transmission, but needs the traversal calculation to find the destination tag with hash function, and doesn't resist the asynchronous attack. The location index of the tags uses a plaintext in the Chen Shaowei's protocol, which is vulnerable to the tracking attack and denial of service attack. Zhou [14] introduces the pseudo random number based on Hash-chain, and the update cost of the key is large. Liu Peng et al. make use of the random numbers produced by the reader and tag, and assemble them with ID of tags as the input of hash function, and transmit the values to the back-end database for calculation and comparison by exhaustive search, but it may lead to poor performance of the system and does not resist DOS attack. The RP and RSP authentication protocols are proposed by Zheng [1]. RSP utilizes the random number generator, the exclusive OR function, the same OR function and hash function respectively to enshroud the interaction information between the tag and reader, and the security is formally proved by using BAN logic, but the transmission of hash value for single ID of the tag will lead to replay attack. The HSASILC protocol for RFID authentication is proposed by Si [17] with GNY logic proving, which introduces the time stamp in each certification step, but it does not resist man-in-the-middle attack.

In short, there are still some problems in the existing RFID authentication protocols based on Hash function, and it is of great practical significance to design efficient, secure and reliable RFID Hash-based protocol with limited cost. So, in this paper a lightweight mutual authentication protocol based on dynamic shared keys is proposed, which is suitable for the RFID system with low cost, low computational cost, and large numbers of Tags.

2 Lightweight RFID Mutual Authentication Protocol Based on Dynamic Shared-Key

In this protocol, the query-and-response mechanism is used and the mutual authentication process is based on the improvement of the storage information in the RFID tags.

2.1 Initial Condition

Initially, the parameters including the location index as ki, identification as $tagID$ and a dynamic shared-key as key are stored in the tag, which is embedded with a Hash function (SHA-1, MD4) and a random number generator. The reader has a random number generator, and the back-end database stores all the records for all tags and readers. A record of a tag should fully include following parameters such as Ki, ID_T, key_{old} and key_{new}, and the backend system can carry out a variety of complex computing. Assume as follows:

$H(x)$ is a one-way Hash function; R_R is a random number generated by the reader, R_{Ti} is a random number generated by the tag; $Rot(A, B)$ realizes circularly the left shift of binary number A with n bit, and n is the binary 1 number of B (Hamming weight). The variables key_{old} and key_{new} own the same value as key by initialization.

2.2 Authentication Steps

The authentication process of this protocol is shown in Fig. 1. The process is specially described as follows:

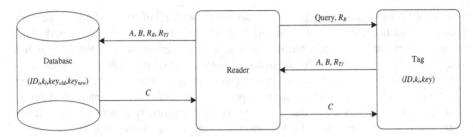

Fig. 1. Certification process of the protocol

(1) The reader, together with the random number R_R generated by itself, sends a Query to the tag as an authentication request.

(2) Within the range of effective communication, there may be more than one tag to respond the reader at the same time, which may lead to the collision of the Radio frequency signal, and cause the failure for tag identification. So the anti-collision protocol will be lunched to ensure that the suitable tag is selected for the response.

(3) The authentication process starts between the selected tag and the reader. The tag generates the random number R_T, and calculates as follows:

$$A = H(Rot(R_T \oplus R_R, R_R)) \oplus k_i \tag{2.1}$$

$$B = Rot(H(ID \oplus key), R_{Ti}) \tag{2.2}$$

where, A is used for encrypting the Ki value to transmit the tag's location index in back-end database, B is used for transmitting the dynamic shared-key secretly. Then the variables A, B and R_{Ti} are sent to the reader by the tag.

(4) The reader will transmit the received variables like A, B and R_{Ti} to the back-end database, as well as the random number RR generated by its own self. The transport way may be through the serial port with wired way or other net way.

(5) The back-end database system receives the information from the reader, and calculates the location index of the record for the response tag in the database.

$$k_i = A \oplus H(Rot(R_{Ti} \oplus R_R, R_R)) \tag{2.3}$$

According to the k_i value the system locates the record of the tag in the database and read the corresponding variables such as ID_i, f_i, key_{old} and key_{new}. If the record search fails or the variable f_i equals to 5 (over 5 times failure), the tag verification is failed and have to turn to (7).

(6) The calculations as following will be done in the back-end database system:

$$B_1 = Rot(H(ID_i \oplus key_{new}), R_{Ti}) \tag{2.4}$$

$$B_2 = Rot(H(ID_i \oplus key_{old}), R_{Ti}) \tag{2.5}$$

If $B_1 = B$ then makes $key_i = key_{new}$ and $f_i = 0$, where, it is to say that both the previous and this authentication are successful;

If $B_2 = B$ then makes $key_i = key_{old}$ and $f_i = 0$, which explains this authentication is successful but the previous is failed;

Otherwise set $key_i = key_{old}$ and $f_i = f_i + 1$, and what is illustrated that both the variables key_{new} and key_{old} in the back-end database is different from the key value in the tag, and both the previous and this authentication are failed. The current tag is considered as illegal.

The calculation will be done as follows:

$$key_{old} = key_i \tag{2.6}$$

$$key_{new} = \text{Rot}(ID_i \oplus key_{old}, R_R \oplus R_{Ti}) \tag{2.7}$$

$$C = \text{H}(\text{Rot}(ID_i \oplus key_{old}, R_{Ti})) \tag{2.8}$$

The system updates the k_i-th record of database with parameters f_i, key_{old} and key_{new}.

If $B = B_1$ or $B = B_2$ then turn to (8).

(7) Set

$$C = \text{H}(ID_i \oplus R_{Ti} \oplus R_R) \oplus f_i \tag{2.9}$$

(8) The back-end database transfers the C value to the reader, which forwards it to the tag.

(9) The tag receives the C value from the reader and then calculates:

$$C_1 = \text{H}(\text{Rot}(ID \oplus key, R_{Ti})) \tag{2.10}$$

If $C_1 = C$ then the reader and tag are legitimate, and the key value will be calculated and updated in the tag:

$$key = \text{Rot}(ID \oplus key, R_R \oplus R_{Ti}) \tag{2.11}$$

Otherwise the authentication fails, the agreement is terminated.

2.3 Protocol Characteristics

This protocol has the following characteristics:

(1) Hide the location index the tag ID in the database to avoid the exhaustive search for each tag ID and comparison;

(2) Hide the authentication key through the transformation of the hash value of the communication;

(3) Record the last two certified keys in the database, and the key_{old} is the final key;

(4) Record the number of failures to prevent unrestricted attacks for authentication;

(5) Support the ownership transfer of tags. After the tag and reader finish the mutual authentication, the key value shared by the tag and back-end database is updated dynamically, and can be normally used after the tag ownership is transferred;

(6) Support the system scalability. Increase or decrease in the number of tags will not significantly affect the system performance.

3 Safety and Performance Analysis

3.1 Security Analysis

(1) Confidentiality. Since the tag interior is safe, it is difficult to obtain the internal key and the identifier of tag unless the attacker makes the reverse engineering analysis of the tag's internal circuit. Even though the current session information of the tag Ti is known, the attacker can not obtain the tag ID because the communication information only includes A, B, C, R_R and R_{Ti} between reader and tag, which are packed by using the unidirectional Hash function except the random numbers, so the protocol can guarantee the anonymity of the tag. After each successful authentication, the shared key in the tag and database is synchronously updated. For each authentication request, the tag responses include the A and B values are calculated by using the shared key and random numbers of a new round, as well as the Hash function. So each response from the tag to reader is not the same, i.e. the tag has the indistinguishability.

(2) Integrity. All the received datum will be calculated and verified by use of the one-way characteristic of Hash function, any modification on the data will lead to the failure of the authentication, which can guarantee the integrity of the data.

(3) Forward security. Each authentication request makes use of a random number of new round to calculate the A and B values. The tag ID only can be used by the tag own, thus an attacker is unable to figure out the tag ID from the hash value, and cannot work out the last key to decrypt the last message from this key value yet. So the attacker does not recognize the last session of the Ti tag, and it's past behavior cannot be traced.

(4) Backward security. The each response of A and B values from the tag are worked out of the random number R_R and R_{Ti} by hash function and Rot-function in tag. The shared-key between the tag and back-end database is calculated with the key (key_{old}) of current session and hash function for update, and the attacker is unable to get the update parameters of the key only by eavesdropping. In the case of R_R and R_{Ti}, the attacker cannot obtain the key information needed for the next authentication by self-calculation.

(5) Anti replay attack. The tag and the reader respectively have new random numbers in each certification process. These random numbers ensure the freshness of the transport message for the authentication based on challenge-response mode, and each successful authentication makes the new shared key updated synchronously between the tag and the database. Therefore, though the attacker repeatedly sends authentication request to the tag with the same random number R_R, the responses will be different by hash encryption, and the different tags response different messages because of different random numbers, so the tag will not be tracked.

(6) Anti desynchronization attack. The shared-key value in the tag will not be updated because of unsuccessful authentication, but the key_{old} in the back-end database all the time keeps the shared-key value for the successful authentication right, which can make sure of using the right key in the next authentication. So the synchronization of secret information can be kept between the server and the tags.

(7) Anti DOS attacks. This protocol does not limit the number of access tags instead of the number of failed authentication. If the third session between reader and tag in the certification process is blocked, that leads to the dynamic shared-key in back-end database updated but the corresponding key in tag not updated synchronously. However, $key_{old} = key$ in the database will be the right key for the authentication next time, the updated key value keynew will be invalid. While the reader launches the next authentication, the equation $key = key_{old}$ in the back-end database is still set up, the tag can still be certified. So the protocol has a good resistance to denial of service(DOS) attacks.

According to the security of seven aspects: indistinguishability, forward security, replay attack, spoofing attack, non traceability, can not track of key, dynamic Key update, and anti desynchronization attacks, the proposed protocol is compared with Hash-Lock protocol (HL), Random-Hash-Lock protocol (RHL), Hash-Chain protocol (HC) and the two improved protocol in 12th reference (Ref. 12) and 1st reference (Ref. 1). By comparison as shown in Table 1, it is found that the proposed protocol has better security than other protocols.

Table 1. Comparison of the security of the protocols

Security	HL	RHL	HC	Ref. 12	Ref. 1	This protocol
Indistinguishability	×	✓	✓	✓	✓	✓
Forward security	✓	✓	✓	✓	✓	✓
Replay attack	×	×	×	✓	×	✓
Spoofing attack	×	×	×	✓	✓	✓
Non traceability	×	×	✓	✓	✓	✓
Dynamic key update	×	×	×	×	×	✓
Anti desynchronization attack	O	O	O	×	O	✓

The protocol has security with the case of: X: does not; ✓: has; O: leaves out of account.

3.2 Computational Performance Analysis

The Hash value and the shared key are required to calculate in the tag and the back-end database for Hash-based RFID authentication protocol, but the storage capacity and the amount of computation will affect the efficiency of the implementation of the protocol and the production cost of the tags. The performance of each protocol is analyzed from two aspects: the calculation amount and the storage capacity of the tag and the back-end database. Comparisons are as shown in Tables 2 and 3, where N denotes the number of tags, H says hash function, L shows logic operations, M figures Hash-chain length, $O(x)$ is the complexity of the calculation for searching tags in back-end database.

Table 2. Comparison of calculation

Calculation	HL	RHL	HC	Ref. 12	Ref. 1	This Protocol
Backend DB	$o(1)H$	$o(N)H$	$o(MN)H$	$o(N)H$	$o(1)H + o(1)L$	$o(1)H + o(1)L$
Tag	H	H	$2H$	$3H$	$2H$	H

Table 3. Comparison of storage capacity

Storage cap.	HL	RHL	HC	Ref. 12	Ref. 1	This protocol
Backend DB	$4l * N$	$l * N$	$2l * N$	$2l * N$	$3l * N$	$4l * N$
Tag	$2l$	l	l	$3l$	$3l$	$3l$

As can be seen from Table 2, compared with other protocols, the amount of computation for this protocol no matter on the tag or in the back-end database is correspondingly less. So it improves the efficiency of the authentication. However, as Table 3 shown, the storage capacity of this protocol in tags and back-end database storage is a bit more (where l is the length of the shared key) than the others, which has almost no impact on the calculation.

3.3 Proof of the Protocol with SVO Logic

The security of this protocol is proved by SVO logic [21, 22]. SVO logic is proposed by Syverson and Van Oorshot, which is optimized and derived from four kinds of logics including BAN, GNY, AT and VO. With very simple inference rules and axioms, SVO Logic repairs the defects and deficiencies of other Logics like BAN.

In the course of proof, R represents the reader (with database), T represents the tag. The axiom A1, A2, A3, A4 are shown as in the references [21, 22]. During SVO logical reasoning, those symbols "$\equiv$", "$\triangleleft$", "$|\sim$", "$|\approx$", "$|\Rightarrow$", "$\ni$", "#" and "$\equiv$" are still used to express "believe", "received", "said", "say", "control", "has", "fresh" and "equivalent" respectively. The analysis of the RFID mutual authentication protocol is as follows:

1. Initial hypothesis

    P1: $R \mid \equiv \#R_R$, $T \mid \equiv \#R_{Ti}$

    P2: $R \mid \equiv R \ni K$, $R \mid \equiv R \ni ID$, $T \mid \equiv T \ni ID$, $T \mid \equiv T \ni K$

    P3: $T \mid \equiv T \overset{K}{\leftrightarrow} R$, $R \mid \equiv T \overset{K}{\leftrightarrow} R$

    P4: $T \mid \equiv ((K, ID, R_R, R_{Ti}) \mid \Rightarrow (A, B, C))$, $R \mid \equiv ((K, ID, R_R, R_{Ti}) \mid \Rightarrow (A, B, C))$

    P5: $T \triangleleft R_R$, $T \triangleleft C$

    P6: $R \triangleleft (A, B, R_{Ti})$

    P7: $R \mid \equiv R \triangleleft *1$, $T \mid \equiv T \triangleleft *2$ (An understanding of the received message by the subject, unknown message)

    P8: $R \mid \equiv (R \triangleleft *1 \supset R \triangleleft (A, B, R_{Ti}))$ (Interpretation of the received messages by the subject)

    P9: $T \mid \equiv (T \triangleleft *2 \supset T \triangleleft C)$

P10: $R \mid \equiv (R \lhd \{X^T\}_K \wedge R \overset{K}{\leftrightarrow} T \supset T \mid :X)$

P11: $T \mid \equiv (T \lhd \{X^R\}_K \wedge R \overset{K}{\leftrightarrow} T \supset R \mid :X)$

2. Proof goal

    G1: $R \mid \equiv (T \ni K)$
    G2: $R \mid \equiv (T \ni ID)$
    G3: $T \mid \equiv (R \ni K)$
    G4: $T \mid \equiv (R \ni ID)$
    G5: $R \mid \equiv \#R_{Ti}$
    G6: $T \mid \equiv \#R_R$

3. Derivation by using SVO Logic

SVO logic has 20 axioms and 2 derivation rules, see References [21, 22]. NEC rule is that $\mid -P \mid \equiv \Phi$ can be derived by $\mid -\Phi$; MP rule is that ψ can be derived by Φ and $\Phi \supset \psi$.

Firstly, an inference can be made by P6, P8 and Trust axiom $(P \mid \equiv \varphi \wedge P \mid \equiv (\varphi \supset \psi) \supset P \mid \equiv \psi$, which is denoted by A1 in this paper):

$$R \mid \equiv R \lhd (A, B, R_{Ti}) \square \tag{3.1}$$

Secondly, an inference can be made by P3, P10, formula (3.1) and A1:

$$R \mid \equiv T \mid :\{A, B, R_{Ti}\}_K \tag{3.2}$$

So the formula "$R \mid \equiv (T \ni K)$" is established, and the goal G1 has to be permitted.

The following formula can be deduced by P1, P4, A1, NEC rule and Message-freshness axiom $(\#(Xi) \supset \#(X1, X2, ..., Xn)$, which is denoted by A2 in this paper):

$$R \mid \equiv \#\{A, B, R_{Ti}\}_K \tag{3.3}$$

It can be reasoned out from the formulae (3.2) and (3.3), the rule NEC and the temporary-value-verification axioms $((\#(Xi) \wedge P \mid :X) \supset P \mid \approx X$, which is denoted by A3 in this paper):

$$R \mid \equiv T \mid \approx \{A, B, R_{Ti}\}_K \tag{3.4}$$

Furthermore, the inference can be worked out by the formula (3.2) and the message sending axiom $(P \mid \approx (X1, X2, ..., Xn) \supset P \mid : (X1, X2, ..., Xn) \wedge P \ni Xi$, which is denoted by A1 in this paper):

$$R \mid \equiv (T \ni (A, B, R_{Ti})) \tag{3.5}$$

So the formula "$R|\equiv(T \ni ID)$" is established according to the formula (3.5), P4, A1 and the message understanding axiom $(P|\equiv(P \ni F(X)) \supset P|\equiv(P \ni X)$, which is denoted by A5 in this paper), and the goal G2 gets permit.

In succession, the inference can be made by P5, P7, P9 and A1 as follows:

$$T|\equiv T \lhd \{C\}_K \tag{3.6}$$

It can be easily inferred out by P3, P11, the formula (3.6) and A1 as follows:

$$T|\equiv R|{:}\{C\}_K \tag{3.7}$$

So the formula "$T|\equiv(R \ni K)$" is set up, and the goal G3 gets permit.

Similarly, it can be deduced by the P1, A2, A1 and NEC rule as follows:

$$T|\equiv \#\{C\}_K \tag{3.8}$$

The following formula can be reasoned out from the formulae (3.7) and (3.8), A3, A1and NEC rule:

$$T|\equiv R|\approx\{C\}_K \tag{3.9}$$

An inference can be made by the formula (3.9), A4, A1and NEC rule as follows:

$$T|\equiv(R \ni C) \tag{3.10}$$

So the formula "$T|\equiv(R \ni ID)$" is established and the goal G4 gets permit.

The formula "$R|\equiv\#R_{Ti}$" can be referred out by the formulae (3.4) and (3.5), A2 and A1, therefore the goal G5 gets permit.

The formula "$T|\equiv\#R_R$" can be referred out by the formulae (3.9) and (3.10), A2 and A1, therefore the goal G6 gets permit.

The formal proof of G1 to G6 shows that after successful implementation of this protocol, reader R and tag T with its ID would both trust the shared-key between them. Furthermore, the tag T trusts the random number RR which is sent by the reader is fresh, and the reader R trusts that the random number RTi which is sent by the tag is fresh.

4   Conclusions

RFID authentication protocol is the key guarantee for the safe and stable operation of RFID system. In the light of the analysis of the Hash-based RFID authentication protocols and those improved protocols, a novel lightweight RFID mutual authentication protocol based on Hash function is proposed, and the SVO logic verification and performance analysis of the protocol are carried out. The new protocol uses the random numbers and the hash function to transfer secret authentication information with a limited number for invalid authentication. Compared with the existing protocols, it supports ownership transfer and quantity scalability of Tags, and has the characteristics of resisting spoofing attack, replay attack, tracking attack, anti asynchronous attack and privacy protection. So it offers good security and high application value. However, storage space of the Tag in the new protocol is slightly larger, and the computational load on the tag side will further be reduced so as to reduce costs of tags in the future work.

Acknowledgments. This research is supported by Natural Science Foundation of China (NSFC), under grant number 61370227, and by Union NSF of Hunan Province & Xiangtan City of China, under grant number: 2015JJ5034.

References

1. Zheng, Z., Mo, H.: Research and implication of RFID security authentication protocol. Master dissertation, Beijing Jiaotong University, Beijing, April 2014
2. Sun, X., Zhao, Z.: A Hash-based mutual authentication protocol for the RFID system. J. Hangzhou Dianzi Univ. **32**(6), 29–32 (2012)
3. Sarma, S.E., Weis, S.A., Engels, D.W.: RFID systems and security and privacy implications. In: Kaliski, B.S., Koç, ç.K., Paar, C. (eds.) CHES 2002. LNCS, vol. 2523, pp. 454–469. Springer, Heidelberg (2003). doi:10.1007/3-540-36400-5_33
4. Weis, S.A., Sarma, S.E., Rivest, R.L., Engels, D.W.: Security and privacy aspects of low-cost radio frequency identification systems. In: Hutter, D., Müller, G., Stephan, W., Ullmann, M. (eds.) Security in Pervasive Computing. LNCS, vol. 2802, pp. 201–212. Springer, Heidelberg (2004). doi:10.1007/978-3-540-39881-3_18
5. Ohkubo, D., Suzuki, K., Kinoshita, S.: Hash-chain based forward-secure privacy protection scheme for low-cost RFID. In: Proceedings of the 2004 Symposium on Cryptography and Information Security (SCIS 2004), Sendai, pp. 719–724 (2004)
6. Henrici, D.,Muller, P.: Hash-based enhancement of location privacy for radio-frequency identification devices using varying identifiers. In: Proceedings of the Second IEEE Annual Conference on Pervasive Computing and Communications Workshops, pp. 149–153. IEEE (2004)
7. Rhee, K., Kwak, J., Kim, S., Won, D.: Challenge-response based RFID authentication protocol for distributed database environment. In: Hutter, D., Ullmann, M. (eds.) SPC 2005. LNCS, vol. 3450, pp. 70–84. Springer, Heidelberg (2005). doi:10.1007/978-3-540-32004-3_9
8. Molnar, D., Wagner, D.: Privacy and security in library RFID: issues, practices, and architectures. In: Proceedings of the 11th ACM Conference on Computer and Communications Security, Washington, DC, pp. 210–219 (2004)
9. Tsudik, G.: YA-TRAP: yet another trivial RFID authentication protocol. In: Fourth Annual IEEE International Conference on Pervasive Computing and Communications Workshops. PerCom Workshops 2006, pp. 640–643. IEEE (2006)
10. Li, Z., Lu, G., Xin, Y.W.: A extensible authentication protocol based on Hash chain. Comput. Eng. **34**(4), 173–175 (2008)
11. Xiong, W., Xue, K., Hong, P., et al.: A RFID security protocol based on Hash chain in two-dimensional interval. J. China Univ. Sci. Technol. **41**(007), 594–598 (2011)
12. Yuan, S.-G., Dai, H.-Y., Lai, S.-L.: Hash-based RFID authentication protocol. Comput. Eng. **34**(12), 141–143 (2008)

13. Chen, S., Chen, R., Ling, L.: An improved Hash-function security protocol for RFID bidirectional authentication. Comput. Syst. Appl. **19**(3), 67–70 (2010)
14. Zhou, Y.: Research on RFID mutual authentication protocol based on Hash chain. Master dissertation, South West Jiaotong University (2012)
15. Liu, P., Zhang, C., Ou, Q.Y.: A Hash-based f mutual authentication security protocol for the mobile RFID. Design. Comput. Appl. **33**(5), 1350–1352 (2013)
16. Ding, Z., Li, J., Feng, B.: Research on Hash-based RFID security authentication protocol. J. Comput. Res. Dev. **46**(4), 583–592 (2009)
17. Si, C., Wen, G.: A design and implementation of RFID security authentication protocol based on Hash function. Master dissertation, University of Electronic Science and technology, Chengdu, December 2013
18. Song, B., Mitchell, C.J.: Scalable RFID security protocols supporting tag ownership transfer. Comput. Commun. **34**(4), 556–566 (2011)
19. Huang, Y.J., Yuan, C.C., Chen, M.K., et al.: Hardware implementation of RFID mutual authentication protocol. IEEE Trans. Ind. Electron. **57**(5), 1573–1582 (2010)
20. Kardas, S., Akgu, M., Kiraz, M.S., et al.: Cryptanalysis of lightweight mutual-authentication and ownership transfer for RFID systems. In: 2011 Workshop on Lightweight Security & Privacy: Devices, Protocols and Applications, pp. 20–25 (2011)
21. Syverson, P.F., van Oorschot, P.C.: On unifying some cryptographic protocol. In: Proceedings of the IEEE 1994 Computer Society Symposium on Security & Privacy. IEEE Computer Society, USA, pp. 14–28 (1994)
22. Syverson, P.F., van Oorschot P.C.: A unified cryptographic protocol logic. Technical report, NRL Publication 5540-227

Photovoltaic Power Prediction Model Based on Parallel Neural Network and Genetic Algorithms

Gaowei Xu and Min Liu[✉]

School of Electronic and Information Engineering, Tongji University,
No. 4800, Caoan Road, Shanghai, China
{0gaowei_xu,lmin}@tongji.edu.cn

Abstract. With the wide application of large-scale photovoltaic systems, photovoltaic power prediction can reduce the negative effects caused by the intermittency and randomness of output power for photovoltaic system. This paper proposes a novel photovoltaic power prediction model based on parallel back propagation neural network (BPNN) and genetic algorithms to predict output power, whose input parameters are historical power output data, historical meteorology data, and meteorology data of the objective day. A parallel BPNN algorithm based on MapReduce is proposed to establish a mapping relationship between input and output through studying large amounts of training sample data. Furthermore, a parallel genetic algorithm based on MapReduce is proposed to optimize BPNN initial weights and thresholds. Experiment results show that the proposed model with parallel BPNN and genetic algorithms can significantly improve prediction accuracy and speed, compared with traditional photovoltaic power prediction model.

Keywords: Photovoltaic system · MapReduce · Neural network · Genetic algorithm · Power prediction

1 Introduction

Facing the increasingly serious problem of global energy shortage and serious environmental pollution, exploring renewable energy sources has gained much attention in recent years. Photovoltaic industry is one of the fastest growing industries of renewable energy, it has become a major electricity source at an extremely rapid pace in several countries all over the world, at least 227 GW of PV are now installed worldwide, 50 GW of solar PV were installed globally in 2015 [1]. Photovoltaic power generation technology has many advantages: clean, reliable, no pollution, no noise and so on. However, due to the randomness and intermittency of PV power generation, it will have a negative effect on power grid when more and more PV power stations connect to large power grids. If PV power output can be predicted accurately, scheduling plan can be adjusted timely to ensure power safe and stable operation of power grid.

With the global continuous development of PV power industry, PV power generation prediction has gradually become a research hotspot in recent years. More and more PV power prediction methods and algorithms have been proposed until now. These methods

© Springer Nature Singapore Pte Ltd. 2017
J. Cao and J. Liu (Eds.): MiPAC 2016, CCIS 686, pp. 99–110, 2017.
DOI: 10.1007/978-981-10-3996-6_8

can be divided into two categories: one is to predict PV power output directly, PV power output is predicted according to meteorology data and historical power output data. The other is predict PV power output indirectly, solar irradiance is predicted based on meteorology data and historical solar irradiance data, then PV power output is calculated by solar irradiance value [2–4]. The techniques used in these methods mainly include multiple linear regression, Markova chain, support vector machine, artificial neural network (ANN) and so on [5–7]. Among them, ANN is the most direct and effective technique for PV power prediction. Furthermore, in ANN-related techniques, BPNN is considered to be the most suitable method for classification and prediction problems. Almonacid F. et al. [3] presented a new methodology for forecasting the output of a PV generator one-hour ahead based on dynamic artificial neural network. Liu J. et al. [4] proposed a novel PV power forecasting model based on BPNN to predict the next 24-h PV power outputs, the proposed model considered aerosol index data as an additional input parameter. Obviously, BPNN is extremely practical and effective in dealing with small-scale dataset. With the arrival of the era of PV big data [8], traditional BPNN algorithm are facing the challenge of large-scale data storage and computation. In order to address the above-described challenges, many correlative study focused on optimizing the neural network weights and thresholds. Alabbas M. et al. [9] used the genetic algorithms to optimize the design of NN architecture in terms of number of hidden layers and the choice of the best parameters (learning rate, momentum term, activation functions). Zhang E. et al. [10] presented a method for the sound quality prediction by using a BPNN based on particle swarm optimization (PSO), which is optimizing the initial weights and thresholds of BPNN through the PSO. Although these approaches have obtained some achievements, they still have a great potential. In recent year, with widely application of parallel and distributed computing technologies, many researchers have attempted to speed up the computation process with parallel computing technologies such as the MapReduce platform. Aljarah I. et al. [11] proposed a parallel MapReduce-based GSO algorithm to speed up the GSO optimization process, experimental results show that the proposed algorithm is appropriate for higher dimensions functions. Xu H. et al. [12] presented a MUSK algorithm based on MapReduce for answering top-k queries over large scale uncertain strings, experimental results showed that MUSK has good scalability over large amounts of data. In conclusion, MapReduce seems tailored for big data jobs.

In this paper, a novel PV power prediction model based on parallel neural network and genetic algorithm is proposed. A parallel BPNN algorithm based on MapReduce programming model can improve its convergence speed and accuracy, a parallel genetic algorithm is developed on BPNN algorithm to improve better converge, both are excellent in dealing with large-scale data. Our experimental results are compared with previous proposed models and illustrate the effectiveness of the proposed model. The rest of the paper is organized as follows: Sect. 2 presents the proposed model and implementations of parallel BPNN and genetic algorithm using the MapReduce model. In Sect. 3, experimental results are presented and analyzed to validate the efficiency of the proposed model. Finally, conclusions are given in Sect. 4.

2   Prediction Model for PV Output Power

PV power output is closely related to multiple meteorological parameters, however, it is difficult to establish the nonlinear relationship between PV power outputs and multiple meteorological parameters with a mathematical function. Therefore, in this section, design details of a parallel BPNN algorithm and a parallel genetic algorithm process are proposed separately. Based on the proposed algorithms, a novel PV power prediction model is proposed.

2.1   Parallel BPNN Algorithm

Back-Propagation Neural Network. Artificial neural networks (ANNs) has been widely applied in many classification and regression problems, BPNN has become one of the most commonly used ANNs in recent years. With the help of error feedback mechanism, BPNN can establish nonlinear input-output mapping relationships without knowing their concrete mathematical formulas. A typical BPNN usually consists of multiple network layers, however, a three-layer BPNN can meet most application requirements. Figure 1 shows the structure of a typical three-layer BPNN, which is composed of one input layer, one hidden layer, and one output layer.

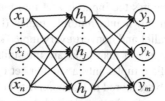

Fig. 1. The structure of a typical three-layer BPNN

The input vector of BPNN is $X = [x_1, x_2, \ldots, x_i, \ldots x_n]$, output vector is $Y = [y_1, y_2, \ldots, y_k, \ldots y_m]$, while the hidden layer vector is $H = [h_1, h_2, \ldots, h_j, \ldots h_l]$. The connection weights $\omega_{i,j}$ between input layer and hidden layer and the connection weights $\omega_{j,k}$ between output layer and hidden layer are usually given randomly.

For hidden layer,

$$h_j = f(\sum_{i=1}^{n} \omega_{i,j} * x_i + \alpha_j) \quad j = 1, 2, \ldots, l$$

The sigmoid function $f(x) = 1/(1 + e^{-x})$ is chosen as the activation function, here α_j is the threshold of jth neuron in hidden layer.

For output layer,

$$y_k = f(\sum_{k=1}^{m} \omega_{j,k} * h_j + \beta_k) \quad k = 1, 2, \ldots, m$$

Here β_k is the threshold of kth neuron in output layer. Afterward, the actual output vector y_k is compared with desired output o_k, the error E is expressed as follows:

$$E = \frac{1}{2} \sum_{k=1}^{m} (o_k - y_k)^2$$

The error E can be decreased continuously until it approaches 0 by adjusting the weights $\omega_{i,j}$ and $\omega_{j,k}$, it is propagated backward from output layer to input layer. Weight updating formula is given as follows:

$$\omega_{i,j}(t+1) = \omega_{i,j}(t) + \eta[(1-\delta)D(t) + \delta D(t-1)]$$
$$\omega_{j,k}(t+1) = \omega_{j,k}(t) + \eta\left[(1-\delta)D'(t) + \delta D'(t-1)\right]$$

η is learning efficiency, $\eta > 0$, $D(t) = -\partial E/\partial w_{i,j}(t)$, $D'(t) = -\partial E/\partial w_{j,k}(t)$, δ is momentum factor, $0 < \delta < 1$.

MapReduce Programming Model. Hadoop MapReduce is a software framework for processing vast amounts of data with a parallel, distributed algorithm on a cluster consist of thousands of computer nodes. It has become the most popular programing model in dealing with large-scale data which is not suitable for running in single computer. A MapReduce job usually separates the input dataset into multiple independent chunks, multiple chunks are broken down into multiple tuples (key/value pairs). Map jobs process these chunks in parallel and convert every tuple (key/value pair) into a set of intermediate tuples. The outputs of map tasks are transmitted to reduce tasks, reduce tasks combines all intermediate tuples into a smaller set of tuples. The input and the output of the MapReduce job are stored in Hadoop Distributed File System (HDFS). A cluster generally consists of one namemode and many datanodes, the MapReduce framework and the Hadoop Distributed File System can run on every node simultaneous. In addition, one namenode is responsible for monitoring tasks, scheduling tasks, and re-executing the failed tasks, while many datanodes execute the tasks assigned by the namemode. Figure 2 shows the MapReduce working principle diagram.

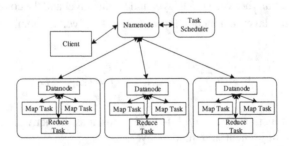

Fig. 2. MapReduce working principle diagram

Algorithm Design. With the arrival of the era of big data, traditional BPNN algorithm cannot meet the real-time requirements, a parallel BPNN algorithm based on

MapReduce programming model is proposed [13]. A training dataset is divided into multiple small datasets, each small dataset is processed by a datanode in a Hadoop cluster for training in parallel, which can greatly improve the convergence speed and accuracy of the BPNN, the framework of parallel BPNN algorithm is given in Fig. 3. When the parallel BPNN algorithm starts, each datanode establishes a BPNN and initializes weights and thresholds, then each datanode reads and processes a data chunk in the form of <key, value> saved on the HDFS. The feedback error of the BPNN is calculated, and then the BPNN adjust weights and thresholds based on feedback error. The BPNN repeat the back-propagation process until all the training data meet the training accuracy requirement. Finally, a namenode collect and merge all the outputs of multiple datanodes.

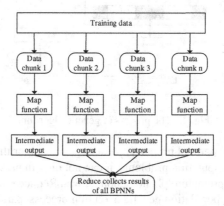

Fig. 3. The framework of parallel BPNN algorithm

2.2 Parallel Genetic Algorithm

Genetic algorithms are important parts of evolutionary algorithms, which are inspired by the principle of nature selection. Genetic algorithms are commonly applied on large-scale optimization and search problems, and can find efficient solutions based on bio-inspired operators within limited time. Figure 4 shows the process of genetic algorithm, genetic algorithm evolution flow are as follows:

Step1: Initialization. Usually, an initial population is generated randomly, the population size M depends on the nature of specific problems. In addition, we set up the evolution algebra counter t as 0, the biggest evolution algebra as T.

Step2: Individual evaluation. A fitness function is used to calculate the fitness value of each potential individual.

Step3: Selection operators. The quality of each potential individual depends on its fitness value, better individuals are selected to the next generation with higher probability in the iterating process.

Step4: Crossover operators. Create new individuals by selecting and recombining genes from a pair of individuals in the current generation.

Step5: Mutation operators. A single individual in the current generation is changed randomly to create a new individual. Then, the fitness value of all offspring are evaluated.

Step6: Termination. Repeat steps 3–5 until a termination condition has been reached.

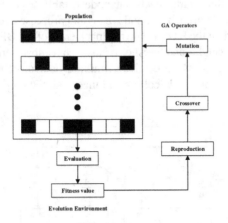

Fig. 4. The process of genetic algorithm

Due to the intrinsic parallel property, parallel genetic algorithm is suitable to solve large-scale parallel computation problems, it can not only improve the speed of operation, but also prevent premature phenomenon. A MapReduce operation is invoked for each generation of the population genetic evolution process, parallel genetic algorithm based on MapReduce is implemented by map operation and reduce operation. To begin, the input data set is split into several data blocks. Each data block is processed by an independent map operation, map operation evaluate the fitness of each individual. The results of each map operation are used as input of reduce operation. Reduce operation can implement selection, crossover, mutation and so on. Hence, the flow chart of the parallel genetic algorithm is shown in Fig. 5.

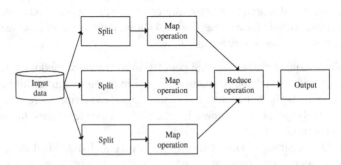

Fig. 5. The flow chart of the parallel genetic algorithm

From the above analysis, Map function reads the individual groups saved on the HDFS, and then evaluates the fitness of each individual, finally, the best individual can be selected and wrote into HDFS. Map function pseudo code is as follows.

```
Map(key, individuals)
{
        For individual in individuals
            fitness=CalculateFitness(individual);
            value=bind(individual,fitness);
            If meet the migration condition
                    key=ChangeKey(key);
            end if
            emitIntermediate(key,value);
        end for
}
```

Reduce function reads the output of map function saved on the HDFS, and combines the values with the same key. Then, the selection, crossover and mutation operation of sub populations are realized. Finally, sub population is wrote into HDFS and used as the input data of the next MapReduce operation. Reduce function pseudo code is as follows.

```
Reduce(key,values)
{
        for value in values
            Individuals[i]=value.Individual;
            Fitness[i]=value.fitness;
        end for
        Individuals=Selection(individuals);
        Individual=Crossover(individuals);
        emitIntermediate(key, Individuals);
}
```

2.3   PV Power Prediction Model

Facing with the fact of bursting data in the PV industry, it is necessary and feasible to introduce the parallel computing technology to PV power prediction model. On the basis of parallel BPNN algorithm and genetic algorithm, a novel PV power prediction model is proposed, its diagram is shown in Fig. 6 [4]. It is well-known that the PV power output mainly relates to meteorological factors, such as temperature, humidity, solar irradiance, aerosol index and so on. Furthermore, meaning and specification of the input/output variables of the proposed model are described in Table 1. In this study, the parallel BPNN can establish the relationship between PV power output and meteorological factors. In addition, in order to improve the convergence speed and accuracy of the BPNN, the initial weights and thresholds are optimized by the parallel genetic algorithm. According to the historical PV power output and historical meteorology data, PV power prediction model is established to fit training. With the well-trained BPNN, hourly PV power output of the objective day can be predicted by inputting the meteorological data of the objective day.

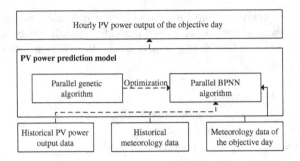

Fig. 6. PV Power prediction model diagram

Table 1. The input/output variables of the proposed model

Input/output variables	Description
x_1	Temperature
x_2	Humidity
x_3	Solar irradiance
x_4	Wind speed
x_5	Aerosol index
$x_6 - x_{17}$	Historical output power
$o_1 - o_{12}$	12 h PV power output

3   Experimental Results

In this section, a comparison of the proposed model and traditional BPNN model was made to evaluate the high efficiency and scalability of the proposed model. After performance evaluation, a case study is given to validate the feasibility and practicality of the proposed model through its application to an actual PV power station. All the experimental results and analysis are shown as follows.

3.1   Experiment Setup and Dataset

We have implemented parallel BPNN algorithm and parallel genetic algorithm based on Hadoop. An experimental Hadoop cluster which consist of 6 computer nodes was built to evaluate the performance of the proposed model. Each computer is equipped with the same physical environment as shown in Table 2. The historical PV power system monitoring data and historical meteorological data are acquired from an actual 20-MW capacity PV system located in Shanghai, China, during the period of May 1, 2013 to May 1, 2016, and these data are used as the training set and test set.

Table 2. Computer node physical environment

CPU	Core i7@3 GHz
RAM	8 GB
SSD	1 TB
Network	10/100/1000 MHz Ethernet LAN
OS	Ubuntu 16.02
JDK version	1.7.2, 64 bit
Hadoop version	2.3.2, 64 bit

3.2 Execution Time and Scalability Experiment

Firstly, in order to evaluate that the proposed model based on parallel BPNN algorithm and parallel genetic algorithm is superior to simple BPNN model in dealing with large data set, we altered the number of computer nodes and recorded the corresponding execution time, the execution time is depicted as the curves in Fig. 7.

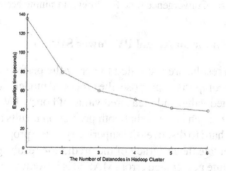

Fig. 7. Execution time for different number of computer nodes

From Fig. 7, as the number of computer nodes increases, the execution time is gradually reduced. Further can see the execution time when running in single computer node is much longer than the execution time when running in multiple computer nodes. When dealing with large scale of data convergence, it becomes very difficult to converge the network during the training process because of insufficient memory. Therefore, the proposed model based on MapReduce programming model is beneficial to improve the convergence speed and accuracy. On the other hand, with the increasing of the number of computer nodes, the proposed model can always implement the convergence and reduce execution time, the scalability performance of the proposed method is obviously.

Secondly, in order to validate that the proposed model obtains higher efficiency with the help of genetic algorithm optimization, we compare convergence time of parallel BPNN algorithm with and without genetic algorithm optimization by changing the training accuracy while the number of computer nodes remains static at 4.

The convergence time of the two methods is depicted as the curves in Fig. 8. The x-axis denotes the different training accuracy and y-axis presents the convergence time of the two methods. As shown in Fig. 8, both two methods seem easier to converge when training accuracy is set to a higher value. Besides, as the convergence accuracy grows from 0.1 to 0.5%, we notice that the convergence time of parallel BPNN algorithm with genetic algorithm optimization is much less than the convergence time of parallel BPNN algorithm without genetic algorithm optimization. To sum up, optimizing BPNN by of genetic algorithm is highly effective.

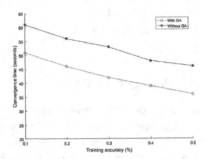

Fig. 8. Convergence time for different training accuracy

3.3   Power Prediction for an Actual PV Power System

The power prediction results are presented to verify the performance of the PV power prediction model. For comparison purpose, the proposed model predicated value, simple BPNN model predicated value and measured value of hourly PV power are depicted in Fig. 9 [4]. From Fig. 9, it can be seen that both prediction results are approximate to the measured values. It is hard to discover the superiority of the proposed model. Therefore, in order to further evaluate the accuracy of the prediction results given by the proposed model, the mean absolute percentage error (MAPE) is introduced to represent prediction accuracy of a prediction method in statistics.

$$MAPE = \frac{100}{N} \sum_{i=1}^{N} \frac{|P_i^p - P_i^m|}{P_i^m} \%$$

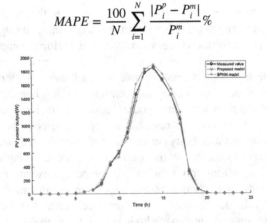

Fig. 9.   PV output power prediction results

The mean absolute percentage error between predicted values P_i^p and measured values P_i^m is calculated by the means of MAPE, the absolute prediction error of the hourly PV outputs is shown in Fig. 10. As it can be found from the results of Figs. 9 and 10, the predicted PV power outputs match measured data well with both simple BPNN algorithm and the proposed model, it is obviously that the proposed model is suitable to realize PV power prediction. Compared with simple BPNN model, prediction results given by the proposed model has a better precision, its average error MAPE is only about 3.53%.

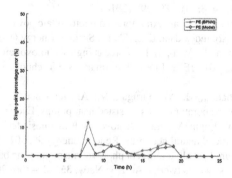

Fig. 10. The absolute prediction error of the hourly PV outputs

4 Conclusion

In this paper, a novel PV power prediction model based on the parallel BPNN algorithm and the parallel genetic algorithm was presented for day-ahead hourly PV power prediction. A parallel BPNN algorithm based on MapReduce has a significant improvement in the prediction accuracy and speed, and a parallel genetic algorithm based on MapReduce is developed for the optimization of neural network initial weights and thresholds, experimental results show that algorithm's execution efficiency is improved enormously by using the parallel processing technique. Furthermore, the results of case study on a practical solar PV power station indicate the effectiveness of the proposed model. Accurate prediction of PV station power generation plays an important role in the safe and stable operation of power grid. As a future work, in order to further improving the prediction accuracy, we plan to pay more attention to classify the PV power output modes under different weather conditions by various data classification algorithms.

Acknowledgement. The research work presented in this paper is partially supported by the Scientific Research Projects of the NSFC (Grant No. 61173015, 61573257) and Hangzhou Municipal Science and Technology Bureau of social development and scientific research projects (No. 20150533B16).

References

1. Sahu, B.K.: A study on global solar PV energy developments and policies with special focus on the top ten solar PV power producing countries. Renew. Sustain. Energy Rev. **43**, 621–634 (2015)
2. Yang, H.T., Huang, C.M., Huang, Y.C., et al.: A weather-based hybrid method for 1-day ahead hourly forecasting of PV power output. IEEE Trans. Sustain. Energy **5**(3), 917–926 (2014)
3. Almonacid, F., Pérez-Higueras, P.J., Fernández, E.F., et al.: A methodology based on dynamic artificial neural network for short-term forecasting of the power output of a PV generator. Energy Convers. Manag. **85**(9), 389–398 (2014)
4. Liu, J., Fang, W., Zhang, X., et al.: An improved photovoltaic power forecasting model with the assistance of aerosol index data. IEEE Trans. Sustain. Energy **6**(2), 1–9 (2015)
5. Teo, T.T., Logenthiran, T., Woo, W.L.: Forecasting of photovoltaic power using extreme learning machine. In: 2015 IEEE Innovative Smart Grid Technologies-Asia (ISGT ASIA), pp. 1–6. IEEE (2015)
6. Gunasekar, N., Mohanraj, M., Velmurugan, V.: Artificial neural network modeling of a photovoltaic-thermal evaporator of solar assisted heat pumps. Energy **93**, 908–922 (2015)
7. Chine, W., Mellit, A., Lughi, V., et al.: A novel fault diagnosis technique for photovoltaic systems based on artificial neural networks. Renew. Energy **90**, 501–512 (2016)
8. Hu, T., Zheng, M., Tan, J., et al.: Intelligent photovoltaic monitoring based on solar irradiance big data and wireless sensor networks. Ad Hoc Netw. **35**, 127–136 (2015)
9. Alabbas, M., Jaf, S., Abdullah, A.-H.M.: Optimize BpNN using new breeder genetic algorithm. In: Hassanien, A.E., Shaalan, K., Gaber, T., Azar, A.T., Tolba, M.,F. (eds.) AISI 2016. AISC, vol. 533, pp. 373–382. Springer, Heidelberg (2017). doi:10.1007/978-3-319-48308-5_36
10. Zhang, E., Hou, L., Shen, C., et al.: Sound quality prediction of vehicle interior noise and mathematical modeling using a back propagation neural network (BPNN) based on particle swarm optimization (PSO). Meas. Sci. Technol. **27**(1), 015801 (2015)
11. Aljarah, I., Ludwig, S.A.: A scalable MapReduce-enabled glowworm swarm optimization approach for high dimensional multimodal functions. Int. J. Swarm Intell. Res. (IJSIR) **7**(1), 32–54 (2016)
12. Xu, H., Ding, X., Jin, H., Jiang, W.: Parallel top-k query processing on uncertain strings using MapReduce. In: Renz, M., Shahabi, C., Zhou, X., Cheema, M.A. (eds.) DASFAA 2015. LNCS, vol. 9050, pp. 89–103. Springer, Heidelberg (2015). doi:10.1007/978-3-319-18123-3_6
13. Liu, Y., Yang, J., Huang, Y., et al.: MapReduce based parallel neural networks in enabling large scale machine learning. Comput. Intell. Neurosci. **2**, 1–13 (2015)

Author Index

Printed in the United States
By Bookmasters